Undara Volcano and its Lava Tubes

Looking through The Archway, Undara Lava Tube System.

ANNE AND VERNON ATKINSON

A GEOLOGICAL WONDER OF AUSTRALIA
IN UNDARA VOLCANIC NATIONAL PARK, NORTH QUEENSLAND

In the sixties and seventies a few people came to recognise that the Undara Lava Tubes constitute a geological wonder. Through the persistence of these people, a National Park was gazetted in 1992 so these features are now protected for the enjoyment of all Australians and our overseas guests.

Since my visit to Undara, I fully endorse that gazettal.

The Honourable Molly Robson,
Queensland Minister for Environment and Heritage, 1992-1995.

The National Parks Association of Queensland Inc. initially recommended to the Minister for Local Government and Conservation on 13th February 1967 that the area of the Undara Lava Tubes be declared a National Park.

The Association was pleased when this was eventually achieved in 1992.

The Association commends the publication of this book, which will enable visitors better to appreciate the features and the significance of this important area.

Norman H. Traves,
President.

George Haddock
Hon. Secretary.

ISBN 0 646 20123 9 (Soft Cover) ISBN 0 646 24580 5 (Hard Cover)

Proudly published in Brisbane, Queensland,
by Vernon and Anne Atkinson;

Colour Separations by PEP Colour Pty Ltd;

Printing by Inprint Limited;

Binding by Podlich Enterprises Pty Ltd.

Copies of this book are available from:
Vernon Atkinson, PO Box 505, Ravenshoe,
Q 4872 AUSTRALIA

▸CONTENTS

This book is affectionately dedicated...

to our parents and grandparents,
who gave us the insatiable curiosity
that has so enhanced our lives;

to our family
who have patiently waited for us to return
to a more normal way of life

and

to the memory of our son Tom,
who built the first steps into Barkers Cave
and assisted with our initial exploration.

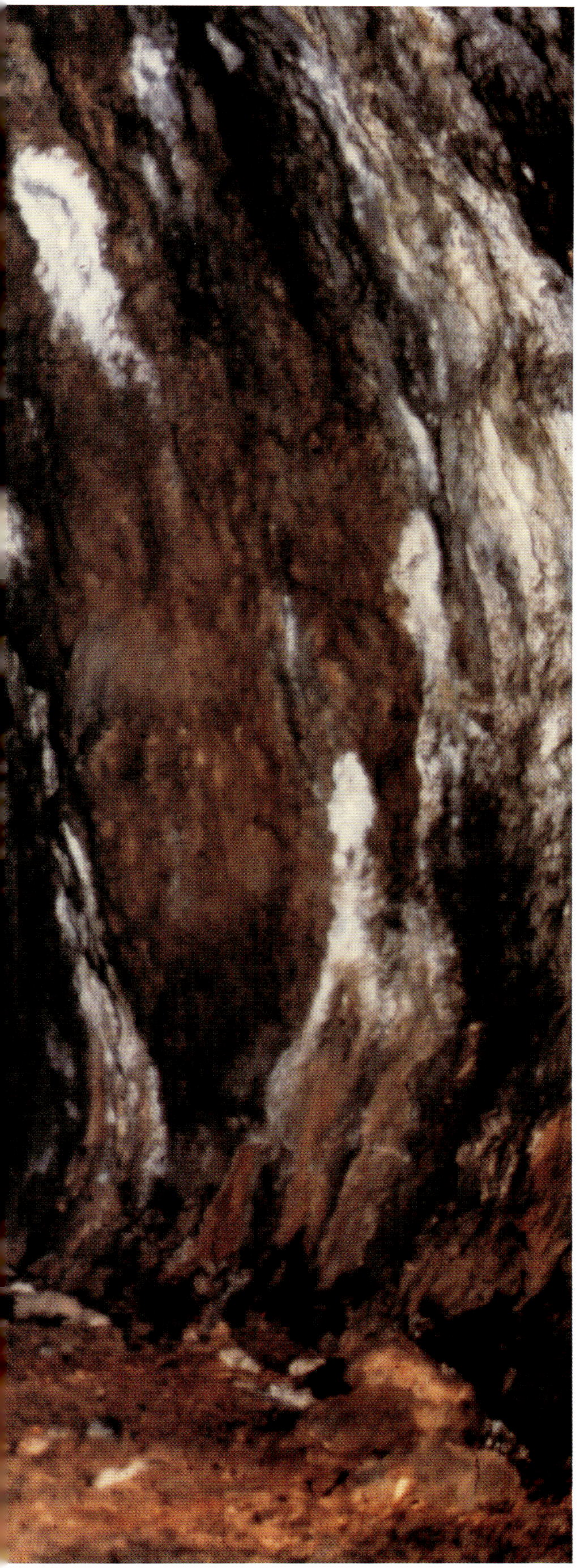

◂ *Barkers Cave, 50 m from the entrance, height 13.5 m. This Cave is 35 km from Undara Volcano, farther than any other cave discovered to date.*
Photo: H.J.L. Lamont, J.C.U.N.Q.

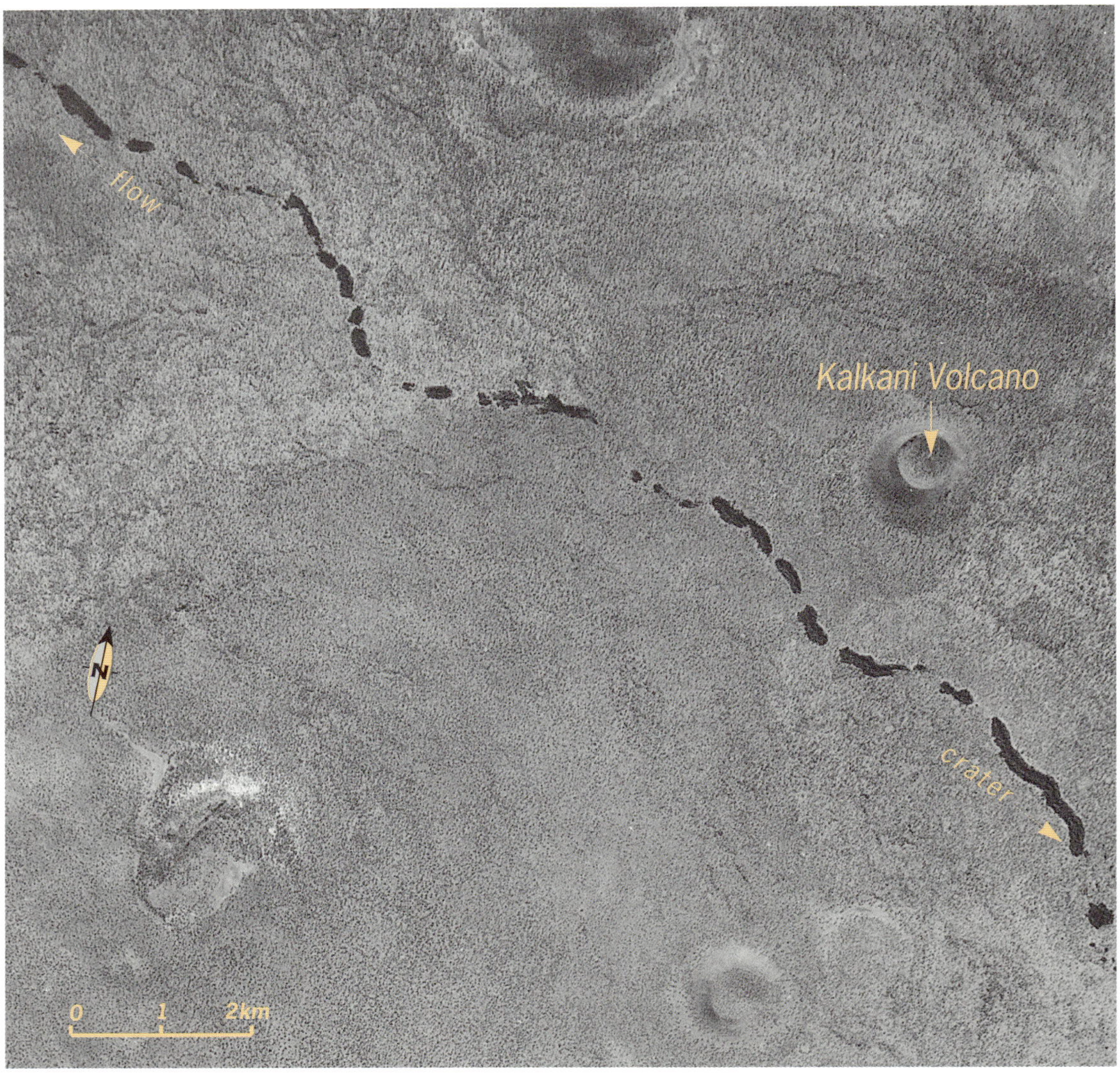

▲
Vertical aerial view of some of the wide depressions aligned with and/or adjacent to the Undara Lava Tube System. Their semi-deciduous rainforest type of vegetation contrasts sharply with the vast open eucalypt forest that surrounds them. The circular crater, marked with an arrow is Kalkani Volcano, a cinder cone that predates the Undara flows.
Photo: by courtesy A.U.S.L.I.G. (see page 76).

▸ FOREWORD

Dick Smith

Photo: Australian Geographic

From the air, Undara's lava tubes can be seen as question marks of dark green vegetation curling through the dry savannah. How many other pilots have looked down and wondered what underground phenomenon could bring about such a curious change on the surface?

Could they have guessed that they were seeing evidence of one of the great wonders of our planet — lava tubes that once channelled molten lava and which are now big enough to drive a road train through and long enough to bring them to the attention of scientists around the world?

Many of these scientists would find it odd that such a fascinating geological formation has come under intense study only in the past two decades. I think this fact alone is fantastic, as it shows that remote Australia still harbours secrets for future generations to reveal.

It's a tribute to Vernon and Anne Atkinson that this book is being published. I've known them as fellow members of the Explorers Club. Their enthusiasm is contagious and, as this book shows, they are more than willing to share what they have learnt.

Dick Smith

Dick Smith
Founder, The Australian Geographic Society.

▲
Mikoshi Arch, Wind Tunnel Complex, Undara Lava Tube System, North Queensland. Visitors with a Savannah Guide from Undara Lava Lodge.
Photo: Kerry Trapnell, by courtesy Undara Experience.

▲
A lava ridge on Venus. It is at least 79 km long and 2–3 km wide and in an area interpreted as basaltic. Many secrets of this planet, long veiled in mist and mystery, have been revealed by analysis of radar data from the Magellan Space Mission, 1990–1992. The ridge, pictured above, is one of the features for which analogues can be found in basaltic areas on Earth. Colour is false — the enhancement is for clarity.
Photo: G. Melville, U. W. Data per courtesy NASA.

◂ *Aerial view of the terminal section of the The Wall, a 35km basalt ridge Northwest of Mount Surprise, North Queensland. This feature is regarded as Earth's best analogue to the smaller basaltic ridges on the Moon and Venus.*
Photo: R. Dann

▸ PREFACE

Initial exploration of the Solar System has shown that volcanism has occurred on nearly all of the terrestrial planets and satellites. Over the last two decades, there has been growing evidence that volcanism of basaltic compositions is present on other planetary bodies, just as it is on Earth. Except for samples limited to the Moon and some meteorites, much of this evidence hinges on the interpretation of remote sensing data and clues derived from the morphology of volcanic features.

Recognising the need to establish the basis for such interpretations, NASA and other space agencies encourage the study of volcanic features that appear to be analogous to those seen on the Moon, Mars, Venus and other planetary bodies. Vernon and Anne Atkinson, their family and colleagues have conducted an exemplary study of the Undara volcanic complex. The lava tubes, both collapsed and uncollapsed, and related flow features provide invaluable insight into the nature of comparable volcanic features on the Moon, Mars and Venus, and enable the development of models for lava flow emplacement. Of particular importance is the great length of the Undara Lava Tube System and the existence of features such as "The Wall".

This book provides information and insight of interest to vulcanologists and lay persons alike. It is superbly illustrated and accurately presented. Of equal importance to the scientific contributions, Anne's perseverance and tenacity in conducting the research serve as an inspiration and model to students and professional colleagues.

Ronald Greeley
Professor of Geology
Arizona State University U.S.A.
Former NASA Research Scientist.

Above: *Moon — Composite.*
Photo: by courtesy University of California, Lick Observatory.

Right: *Astronaut Edwin 'Buzz' Aldrin, pilot of the lunar landing craft, Eagle, on the Moon.*
Photo: by courtesy NASA.

The maps on **this page and opposite,** place Undara Volcano and its lava field in context, geographically.

◂ ***Eastern Australia.***

Extending more than 4,000km from Cape York to Tasmania and within 200–400km (lighter coloured band) of Australia's east coast, remnants of volcanic rock are proof of volcanic activity during the Cainozoic Era. This era, stretching back 66 million years, is not really so long ago, geologically speaking — considering that the age of the Earth is thought to be over 4,500 million years. Boxed area is shown below.

After: P.J. Stephenson, T.J. Griffin and F.L. Sutherland, 1980.

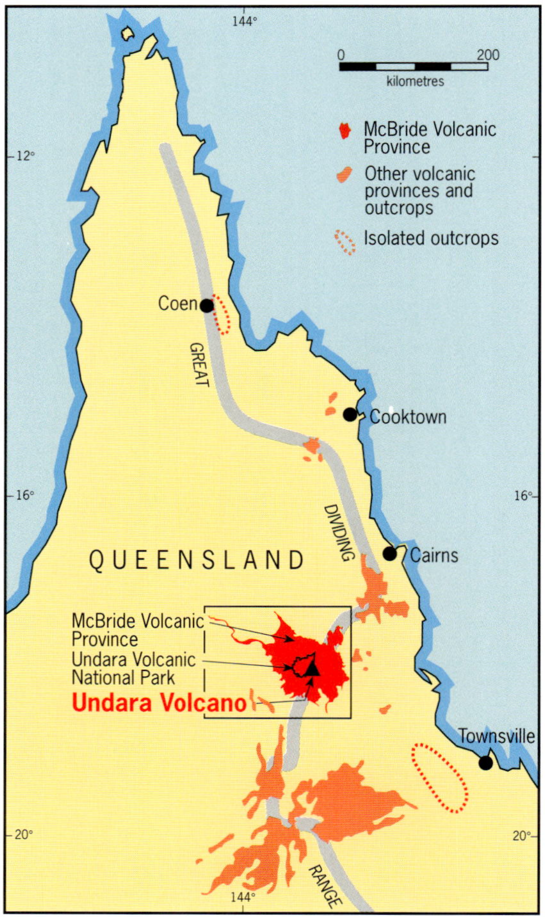

North Queensland.

*In the Cairns–Townsville hinterland, there are five areas (Provinces) of volcanic rock and soil. The **Undara Volcano** is one of more than 160 volcanoes in the McBride Volcanic Province. Boxed area is shown on facing page.* ▸

UNDARA VOLCANO
AND ITS LAVA TUBES

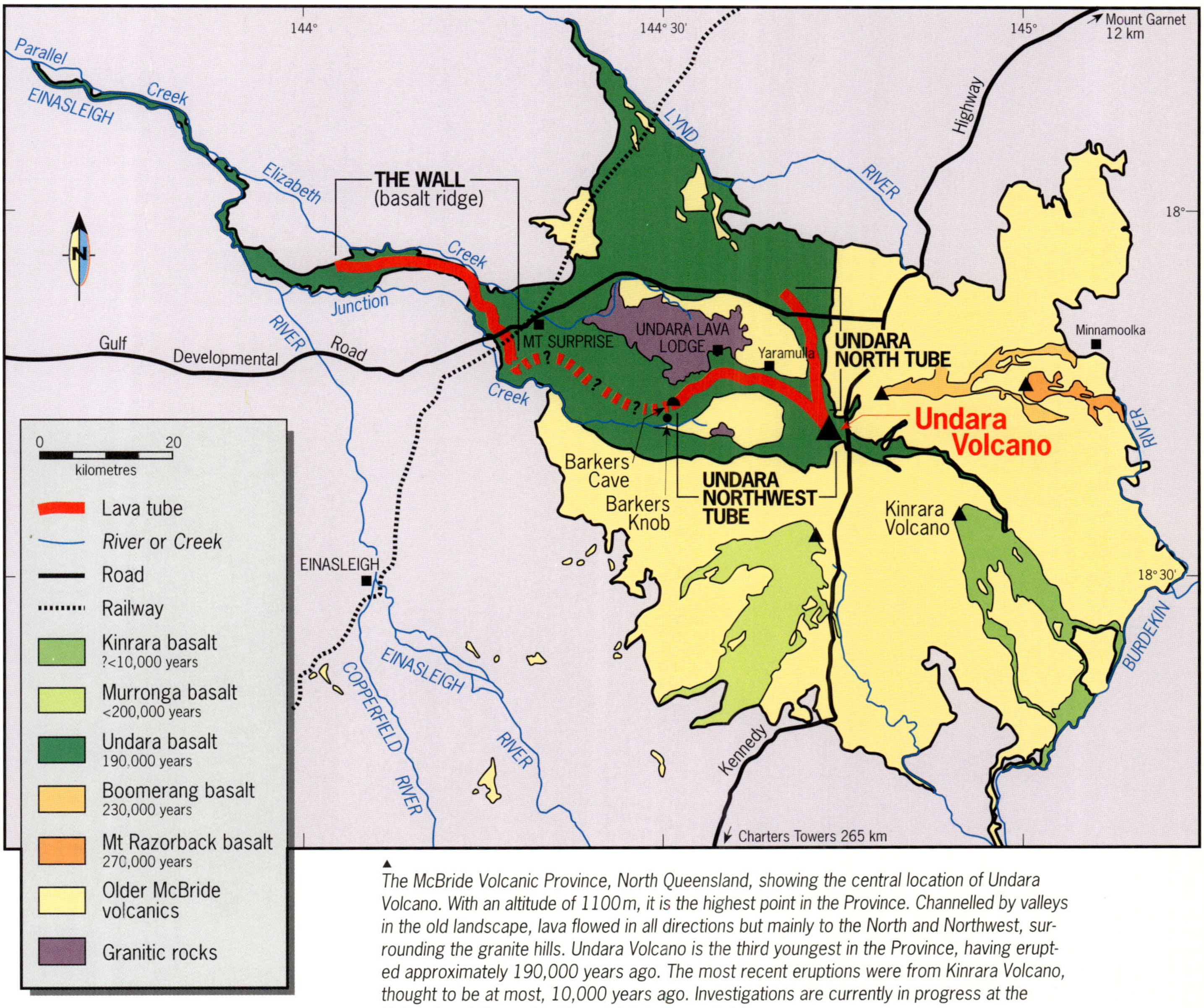

▲

The McBride Volcanic Province, North Queensland, showing the central location of Undara Volcano. With an altitude of 1100 m, it is the highest point in the Province. Channelled by valleys in the old landscape, lava flowed in all directions but mainly to the North and Northwest, surrounding the granite hills. Undara Volcano is the third youngest in the Province, having erupted approximately 190,000 years ago. The most recent eruptions were from Kinrara Volcano, thought to be at most, 10,000 years ago. Investigations are currently in progress at the University of Queensland to establish an exact date.

Map: After T.J. Griffin (1977).

▲
*View of the **Moon** showing a meandering channel, 115 km long, which may represent a collapsed lava tube in a basaltic area. Apollo 15 landing site is indicated by the orange cross. Sun illumination is from the right.*

Photo: Astronaut A.M. Worden, NASA Apollo 15 Mission.
Frame: NASA: AS15–0414. By courtesy NASA.

▸ INTRODUCTION

On the Moon, meandering channels have long puzzled geologists. These channels look like river courses (see photograph opposite) but cannot be, for one simple reason: it has never rained on the Moon because the Moon has no atmosphere.

When astronauts brought back the first rock samples from the Moon, some were of basalt, the most common volcanic rock on Earth, and geologists suggested that the channels on the Moon may be collapsed lava tubes. This new theory, first put forward in 1966, stimulated studies of lava tubes worldwide, including those at Undara, whose tube system is quite remarkable, not only for its geology, but also for its unique fauna and flora.

This book presents a pictorial record of some of the features of Undara's geology. It aims to let you, the reader, share the excitement of discovery there and to have a record of some of the features that you can see and of others that are hidden far from sight.

More than 60 caves and arches have now been discovered in the Undara Lava Tube System. Most are less than 200m long, but the System does include Australia's longest lava tube cave, over 1.3km. More than 6km of caves have been surveyed and the world's first longitudinal profile has been drawn of a source volcano and its representative caves and arches (page 31).

The main branch of the lava tube system extends for more than 110km. It includes an almost level ridge 35km long, known as **The Wall**. This is considered by lunar geologists to be Earth's volcanic feature most like the smaller basaltic ridges on the Moon (photos, page 48).

The authors' initial research at Undara (1972–1974) aimed to discover how the tubes had formed, by studying representative sections near the crater, far from it and midway. The two intermediate sections remained almost unknown until 1988, when young members of London-based Operation Raleigh volunteered to explore them.

A comparison of Undara's tubes with those currently active elsewhere in the world, indicates that they formed by the draining of roofed lava channels whose locations were determined by the topography at the time of the eruptions.

An exciting new discovery of lava tubes has just been made. Analysis of the radar data from NASA's Magellan Space Probe to Planet Venus indicated that, as well as the straight channels in volcanic areas, there are some aligned depressions. It has been proposed that lava tubes connect these depressions... and to explain the origin of several long ridges, recently detected on Venus, we may well be looking to The Wall at Mount Surprise (photos, page viii and 49–51).

The impressive crater of **Undara Volcano** is in a vast area of open eucalypt forest. The crater, 340 m across and 49 m deep, rises only 20 m above its surrounding lava field. It and the aligned depressions (middle distance, top photo) are made conspicuous by dark green rainforest-type vegetation that once covered the region but is now mainly confined to coastal and near-coastal areas.

◂ *Undara Crater viewed from a light aircraft, Dry Season. Some of the aligned rainforest-filled depressions can be seen in the middle distance.*
Photo: Tom Atkinson.

Undara Crater viewed from the north rim, early morning, Winter 1973.
Photo: Anne Atkinson.
▾

▲
Alignment of depressions along the flow.
Photo: H.J.L. Lamont, J.C.U.N.Q.

From Undara Crater, running North and Northwest, are two alignments of depressions up to 100m wide, some elongate, others circular to oval, most with dark green vegetation similar to that of the crater. Early geologists misinterpreted their origin, seeing their alignment as evidence of a volcanic eruption along a fissure. Discussion with geologists of the United States Geological Survey's Hawaiian Volcano Observatory, who observe active lava flows on the Island of Hawaii, indicates that these depressions are drained lava ponds (see pages 46 and 47).

◂ *View across a typical depression, showing the erratic shapes, sizes and arrangement of the blocks of basalt. The rim is near the top of the photograph but somewhat masked by the vegetation.*
Photo: Sarah Collins.

▸ FORMATION OF A LAVA TUBE AND LAVA POND AT UNDARA

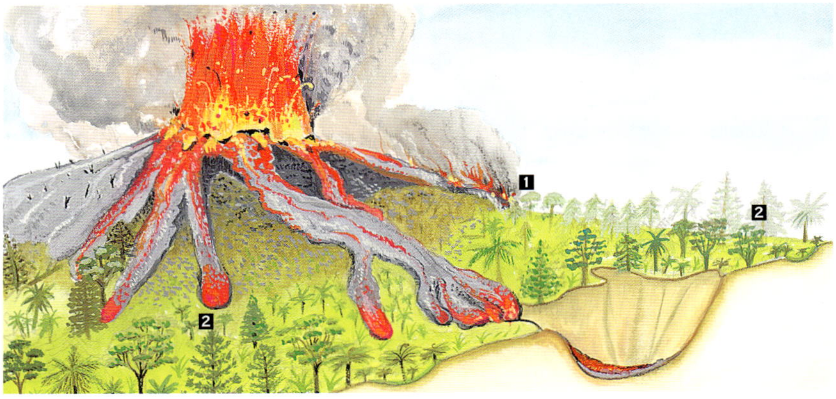

1 *lava flow engulfing trees.*
2 *rainforest-type vegetation.*

About 190,000 years ago — a tremendous volume of treacle-like lava came pouring out of Undara Volcano. The lava blanketed more than 1550 km^2 but parts of the flow were channelled by watercourses, one branch extending 160 km to become the world's longest flow from a single volcano.

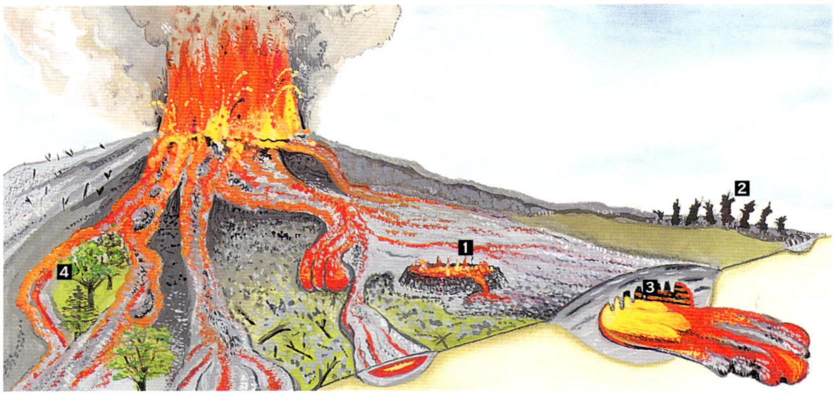

1 *pond of molten lava, retained by its own levée ring.*
2 *tree casts left when the lava flow moved on.*
3 *lavacicles forming as lava drips from tube roof.*
4 *an island of vegetation encircled by lobes of lava — a "kipuka".*

Exposed to the air, the outside of the flow cooled and blackened ... but the lava in the watercourses hardened only on the surface and against the valley walls. Thus a huge insulated conduit formed and lava continued to flow through it.

1 *depression formed when lava pond drained.*
2 *polygonal jointing — the result of cooling and contraction.*
3 *lava level lines — ledges formed by undercutting or deposition whenever flow remained at a constant level.*
4 *floor of cave is the surface of final flow through tube.*

Today — When the eruption waned, most of the lava drained out of the tubes. Where roofs were thin some of them collapsed, providing the entrances that we use today.

Paintings by Sue Livings — original drafts by Anne Atkinson.

▸THE HISTORY OF UNDARA VOLCANO

It is hard for any of us to grasp the geological time scale. To say that the volcanic activity of the McBride Volcanic Province is geologically young seems incomprehensible when it is about 190,000 years since Undara was active ... and, of the other 163 volcanoes still evident in the Province, all but two are older.

An estimated 23 km^3 of lava at about 1200°C erupted from Undara Volcano. Most of the lava was the fluid "pahoehoe" type (see page 65). Flowing like treacle, this bubbling, molten mass spread out across the countryside, blanketing about 1550 km^2.

Some of the lava found its way into old watercourses, mainly to the North and Northwest. Thus channelled, the lava hardened on the top and sides to form insulated tubes which allowed it to flow incredible distances — over 90 km to the North and over 160 km to the Northwest, the latter becoming the world's longest lava flow from a single volcano.

Flowing down a course of gentle gradient, the lava melted sand in the streambed and flowed on and on. Near the granite hill, Barkers Knob, the lava flowed over rapids and a waterfall. The roof was thin and a skylight formed before the flow entered one of the most restricted sections of its entire course. This caused the tube shape to change to a vertical oval with a height of 13.5 m, one of the greatest cave heights measured.

Horizontal, or near horizontal ledges, termed lava level lines, (photo, page 15) on tube walls reflect periods of flow at a constant height in the tube. The width of any ledge reflects how long the lava was flowing at that height.

Surges of lava left multi-layered linings on the tube walls. The number and regularity of these in Pinwill Cave (photo, page 28), known nowhere else in the world, perhaps resulted from proximity to the lava pond immediately upstream — the longest lava pond in the system.

Above the flow, tube walls and roofs became glazed, often re-melting and dripping to form fragile, hollow straw stalagtites and the more robust, triangular lavacicles (photo, page 27) that remain today. Dribbles of lava, like melted candlewax, ran down some walls and occasionally were deflected from vertical by the blast furnace effect of an adjacent skylight, or nearby phreatic explosion, caused by the flow encountering abundant groundwater. At rare locations, drips fell on semi-solid lava and pitted it, like raindrops on a muddy surface.

Lava ponds formed in line with, or adjacent to, some sections of the lava channel. Repeated small overflows and spatter at the margins of ponds built up levées higher than the surrounding country (centre diagram, page 6 and photo, page 46). While the tube was flowing, the ponds remained full, their surfaces de-gassing and crusting as they cooled. When the eruption ceased, the ponds drained back into the tube leaving indescribably untidy depressions lined with blocks, erratic in shape, size and vesicularity (photo, page 5). At a few locations, lava in the ponds drained back into the tube slowly enough to allow small sections of crust to rest, like mosaics, on the inner slopes of the depressions (photo, page 46).

▲
Irregular shaped depressions which resulted from the draining of areas where lava had ponded during the period of active flow.
Photo: Tom Atkinson.

West of Barkers Cave, the lava ponds are irregular in shape, size and location. This may indicate that the lava there had no definite channel and spread widely, possibly over marshy ground.

When the supply of lava finally dwindled, the tubes drained to leave them as we see them today, their floors representing the surface of the final flow that came through them.

Further West, the main flow had entered another major watercourse, the precursor of Elizabeth and Junction Creeks (see map, page 1). Channelled by this for over 40 km, levées built up to form the prominent near-level ridge known as **The Wall** (see pages viii, 49–52), which stands more than 20 m above the surrounding country, Northwest of Mount Surprise.

The toe of the flow hardened — a self-damming effect. The surface of the channel had crusted to considerable depth. As it cooled, it contracted and cracked to form polygonal blocks up to 2 m across. The next surge from the crater is reflected in the slight elevation of the western termination of The Wall (see pages 50 and 52).

A subsequent surge broke through the self-made dam allowing the still molten lava to escape and continue flowing down the watercourse to reach the incredible length of more than 160 km, becoming the world's longest known flow from a single volcano. When the eruption ceased, a central area at the termination of The Wall subsided, leaving a median section sloping gently downward to the West and flanked by wide polygonal columns. About 2 km upflow from the termination, a similar subsidence resulted in a central oval depression — Edmonds Lake — that retains water after rain (see photo, page 52).

The **speed** of the Undara flows can only be estimated from observations of active lava flows overseas. On the Island of Hawaii, the speed of flows in main channels on steep slopes has been measured at up to 56 km per hour. However, when lava from a main channel spreads, it may form flow fronts up to 1 km wide. Because of their great width, flow fronts generally advance at speeds of less than 300 m per hour.

Rate of flow in lava tubes was first measured during the 1970–1971 eruptions on the Island of Hawaii and speeds of between 1 km and 12 km per hour were recorded. Since the eruption of Puu Oo began 11 years ago, lava has continued to flow through some tubes at an average speed of 3.5 km per hour… and so the **time** required for Undara's northwestern flow to reach its final remarkable length of over 160 km, must remain an enigma. Earlier estimates, however, have to be modified in the light of present and recent observations of lava flowing in tubes on the Island of Hawaii.

▸THE MODERN WAY TO VISIT UNDARA COUNTRY...

Photo: A. Little, D.E.H.

Photo: M. Williams, T.C. & K.R.G.

THE MODERN WAY TO VISIT UNDARA COUNTRY...

WITH A SAVANNAH GUIDE FROM UNDARA LAVA LODGE

Photo (above): By courtesy Undara Experience.

Knowledgeable guides will show you maps and, on the way to the caves, point out some of the unique fauna and flora, answer questions and tell you interesting facts about the area.

SAVANNAH GUIDE

When you arrive at a cave the guide will explain the geology, then take you underground (**below**) ... if you wish!

Photo: M. Williams, T.C. & K.R.G.

BUT PREVIOUSLY IT WAS DIFFERENT...

▲
1972 camp for Undara field work. Helium balloons are ready for measuring cave heights. Drafting office (temporary) is to left of Toyota.
Photo: Vernon Atkinson.

▲
1971 — Geologists Stephenson and Griffin make observations from the rim of Kalkani Crater, approximately 12km Northwest of Undara Volcano.
Photo: H.J.L. Lamont, J.C.U.N.Q.

▸THE OPERA HOUSE

This cave was discovered in October 1973, the week that Queen Elizabeth came to Australia to open the Sydney Opera House. Note the "wings" behind the figure.

Photo: H.J.L. Lamont, J.C.U.N.Q.

‣HARBOUR BRIDGE

This arch was named the Harbour Bridge. It formed when two sections of the tube collapsed and left standing an intermediate section 35 m long. In the roof, near horizontal lines mark divisions between surges in the flow (flow units) that came over the original channel crust, thickening and strengthening the roof.

Photo: H.J.L. Lamont, J.C.U.N.Q.

▸ COLLINS ROAD CAVE

This cave has been known for more than 100 years and was named for its proximity to the original waggon road from Herberton. Near the entrance, cool water springs from a crack low on the north wall — no doubt a welcome sight for tired travellers who had bumped over so many miles of rough road.

Sketch by Annette Onslow

▲ *Collins Road Cave, 1995, viewed from the top of the entrance slope. A Savannah Guide from Undara Lava Lodge is explaining the formation to a group of visitors. Note the lines of small rocks. These demark areas, swept once a week and kept free of human tracks to allow daily monitoring of fauna using the cave.*

Photo: Thomas Atkinson.

◂ *Spring water emerging from north wall at the cave entrance.*

Photo: by courtesy Undara Experience.

◂ *Collins Road Cave. On the north wall, near the entrance, lava level lines indicate the varying heights of the lava river when the flow was active. Here they can be viewed in daylight more easily than in any other cave.*
Photo: H.J.L. Lamont, J.C.U..N.Q.

Party leaving after Wet Season visit 1987. The light-coloured vertical stripes are of secondary minerals, leached out by groundwater and re-deposited along cracks in the roof and walls.
Photo: P. Mansbridge, C.C.C.
▾

▸BARKERS CAVE — LOOKING TOWARD THE ENTRANCE

◂ *Entrance to Barkers Cave. The actual scene, viewed from 50 m within the cave.*

Viewed from the same position, with multiple flashes. The steep entrance slope is thought to mark the location of a waterfall or rapids in the pre-basalt landscape. The central arched area is the crust of the last flow through the tube. A final surge beneath this crust caused it to arch and crack longitudinally.

Photos: H.J.L.Lamont, J.C.U.N.Q.

▾

▸BARKERS CAVE — LOOKING DOWNFLOW

◂ *A well drained road? No! This interesting section of elevated floor represents the crust of the final flow in this section of the tube. Fine lava level lines, closely spaced, correspond across the cave on lower walls but do not match within the "gutters". This is seen as evidence that the final flow in this area elevated itself in the centre of the tube by means of minor levées.*
Photo: M. Williams, T.C. & K.R.G.

Cave Coral — a rare and fragile, coral-like structure that is deposited on some cave floors and walls by percolating groundwater. The question is often asked: What is it made of? ...so a tiny piece of this specimen was chemically analysed and found to be calcium carbonate. (Coin for scale is 2.4cm).
Photo: Margaret Atkinson.
▾

Cave entrances are usually inconspicuous (**below**) and represent areas where the tube roof has collapsed. At the entrance to most caves there are only a few vines and minimal rainforest-type vegetation which makes them difficult to detect on air photographs. Closer scrutiny of some entrances may surprise visitors. Occasionally, it is a long, long way down to cave floors (**bottom**), and very dark! Entry to some caves is, in fact, hazardous and requires the skill of an experienced speleologist (**facing page**).

Photo: C.E. Taylor

Where angular blocks of basalt must be negotiated to reach cave floors, people of all ages have been able to view the caves, even those who require a walking stick (**above**). Do not let the boulders daunt you. Tracks have now been marked in caves deemed suitable for visitors.

A typical cave entrance, centre background, in a narrow depression which supports open eucalypt forest, with minimal rainforest-type vegetation at the entrance.
Photo: H.J.L. Lamont, J.C.U.N.Q. ▾

◂ *Looking down into a deep cave entrance.*
Photo: C.E. Taylor.

Facing Page: National Park field officer negotiating overhanging entrance, by flexible metal ladder, during exploration, 1988. ▸
Photo: M.Godwin, D.E.H., C.C.C.

Most cave floors are masked with silt because their entrances are at the higher end, nearer the volcano. This has allowed silt to be washed in.

Pinwill Cave is one of the few that has the entrance at its lower end and there is little silt on its floor.

In Pinwill Cave, an area of floor shows some development of ropy structure (see photo, **below**), a characteristic of modern "pahoehoe" (treacle-like) lava flows. This structure sometimes forms when molten lava under a still-plastic crust, continues to move causing the crust to twist into forms that look like ropes (see photos of ropy lava in recent Hawaiian flows, pages 66 to 68).

Photo: M. Williams, T.C. & K.R.G.

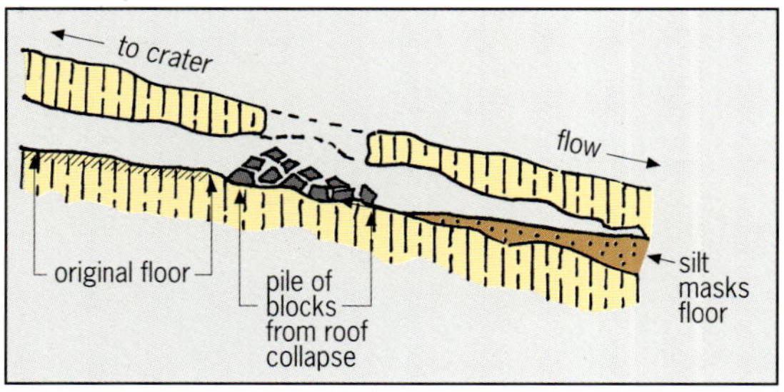

◂ *Diagram to show why so little original cave floor is exposed — less than 200 m of over 6 km surveyed. Most cave entrances are at the end nearer the crater which has allowed silt to wash in and mask floors. It is therefore imperative that we protect what is visible.*

Most caves terminate with a heap of angular blocks of basalt or with a gentle downward curve of roof to floor. Pinwill, Barkers and Ewamin Caves are exceptions. In Pinwill, in addition to fallen rocks at the termination, there is a massive blockage (below) and in Barkers and Ewamin Caves there are lakes (see photos, page 23).

"The Wave" at the base of the massive blockage at the termination of Pinwill Cave. In active flows, waves of lava can be seen rising and breaking like ocean waves. The feature, pictured below, has preserved for all time, the final surge in this part of the tube.

Photo: M. Williams, T.C. & K.R.G.

▲
Roots from surface trees have penetrated parallel roof cracks to form two "curtains" near the termination of Pinwill Cave, an area frequented by variable numbers of bats — numerous females in summer, the bats' "nursery" season.*
**predominant species:* Miniopterus schreibersii.
Photo: H.J.L.Lamont, J.C.U.N.Q.

Below: Two of the unpigmented cave-dwelling invertebrates unique to Undara.
Sketches by Ron Edwards, Kuranda.

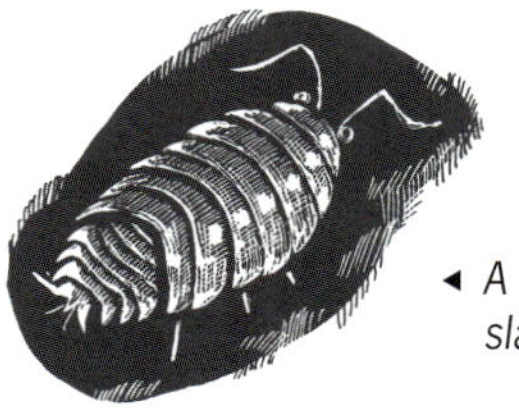

◂ *A scutigerid, or type of slater.*

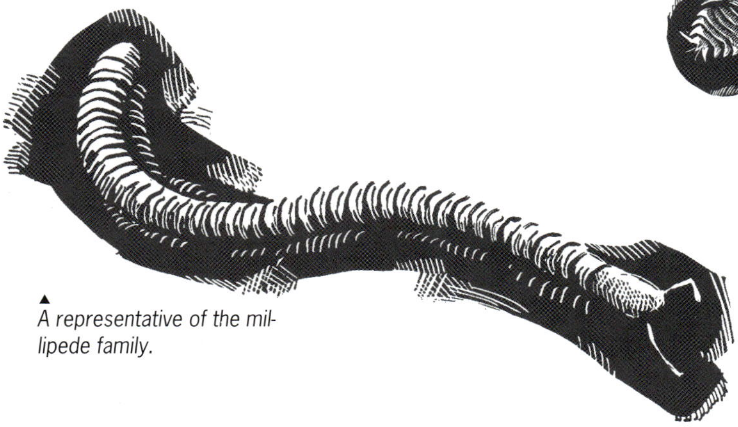

▲
A representative of the millipede family.

▲
The lake in Ewamin Cave in 1991, after a record Wet Season.
Photo: Mark Johnson, by courtesy Undara Experience.

Barkers Cave, about 500m from the entrance. The floor becomes slippery as one approaches a clear still lake.
Photo: M. Williams, T.C. & K.R.G.
▼

— a place for peaceful reflection…?

...but not on this day in 1980 when the Chillagoe Caving Club decided to investigate the extent and depth of the lake!

◂ *Starting to inflate the rubber dinghys.*
Photo: R. Dutton, C.C.C.

Explorers set out.
On return they reported that around the corner, about 150 m on, their journey ended abruptly — the cave roof sloped gently downward to meet the water. On the same day, divers sought an outlet that would give access to further segments of the tube. They could see little through the murky water and found no outlet. They concluded that it must now be filled with silt.
Photo: R. Dutton, C.C.C.
▾

Tube roofs thicken mainly by surges of the flow (flow units) covering the original crust of the lava channel. This is commonly indicated by slight horizontal partings between the layers of basalt in the roof. Ropy lava (see photos, pages 20, 66 to 68), which forms on the surface of flows, is sometimes evident on these layers and there are rare, but significant, imprints of "rope" on cave roofs where blocks have fallen.

Roof structure inside Peterson Cave (east branch). The prominent arched layer just above the observer's head has a ropy surface. Higher ropy surfaces also occur.
Photo: H.J.L. Lamont, J.C.U.N.Q.

Entrance structures of some of the Undara caves. Wavy lines indicate recognisable flow unit surfaces.
Diagram: P.J. Stephenson, J.C.U.N.Q.

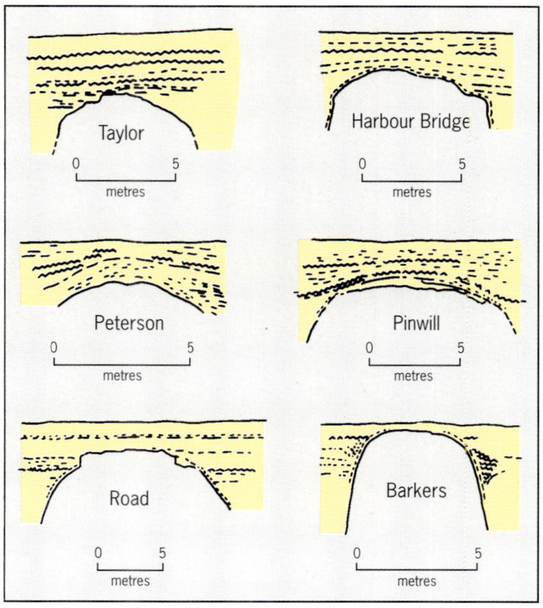

▲
Dribbles of lava on cave walls are usually vertical, as seen in Barkers Cave.

▲
Deflection from vertical is evidence of the speed and heat of the final flow in the tube at this point in Collins Road Cave. This may indicate that there was a "skylight", an opening in the roof of the channel while the lava was flowing, or that the meeting of lava and groundwater caused a phreatic explosion nearby.
Photos: M. Williams, T.C. & K.R.G.

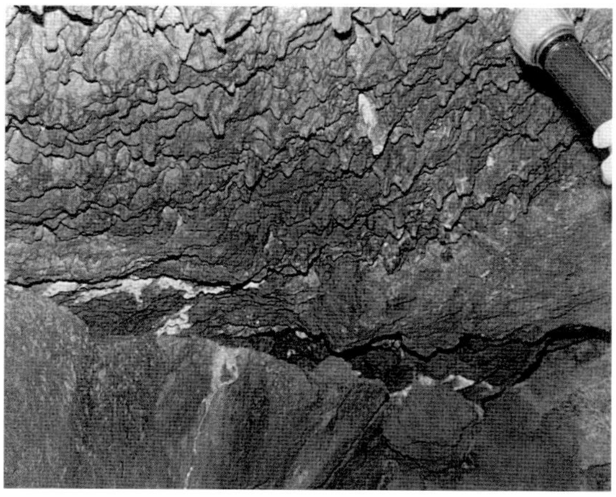

◂ *Lavacicles up to 6 cm long, formed by the dripping of still molten or re-melted lava from roofs and walls.*
Photo: Vernon Atkinson.

◂ *"The Table", a thin sheet of lining near the entrance to Pinwill Cave. The slight curvature may represent a degree of plastic deformation.*
Photo: Vernon Atkinson.

"Rafted" blocks of crust of the final flow have jammed at various angles as slabs of ice jam on a river as it freezes. (Carbide light for scale, height is 32 cm).
Photo: Vernon Atkinson
▾

All caves had linings when first formed. Some linings were a single layer, some multiple. Successive layers did not always anneal to each other and, as in an old home where many wallpapers have been applied, some cave linings have now fallen away to reveal the original tube wall behind them. At rare locations we found linings had curled downward while still plastic.

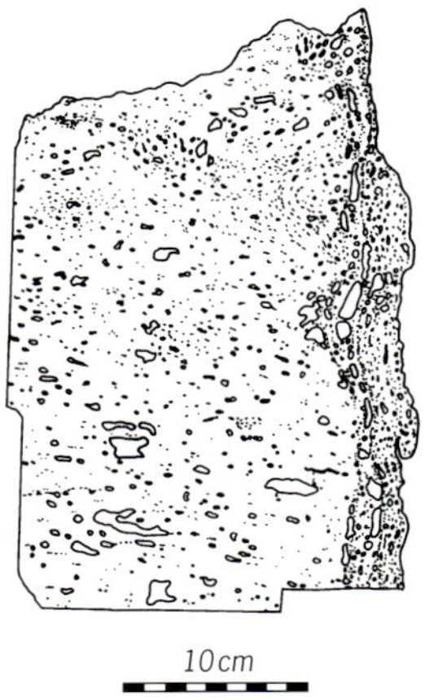

▲
Specimen from the entrance to Barkers Cave. When closely examined, the apparently single lining (right edge) proved to consist of several distinct, annealed layers, indicating deposition of successive linings on still molten surfaces.
Note the elongation of the gas holes (vesicles) and the difference in their alignment: as would be expected, they are predominantly horizontal in the host rock but vertical in the lining. Specimen collected and presented to J.C.U.N.Q by Tom Atkinson.
Diagram: P.J. Stephenson, J.C.U.N.Q.

▲
Multi-layered lining in Pinwill Cave. Up to fifteen layers of lining vary in thickness from 2 cm to 4 cm and represent successive surges of lava through the tube. Nowhere else at Undara, or overseas, have so many layers been recorded. In the foreground are slabs of lining that have fallen from the wall, the flat area which is exposed, upper right. (Torch for scale is 18 cm).
Photo: M. Williams, T.C. & K.R.G.

▲
Tiny blind cave insect, Solonaima baylissa, *collected from a root in Bayliss Cave. Its body size is 4.5mm.*

Photo: by courtesy H. Hoch, P.U., Marburg, Germany.

Right and below: *Roots from surface trees have broken through cave roofs in their search for water. These roots provide food for microscopic and larger cave-dwelling insects.* ▸

Photo right: M. Williams, T.C.& K.R.G.
Photo below: P. Mansbridge, C.C.C.

"The Pavements" in Taylor Cave. These formed where lava movement was slower against the convex channel wall, in a manner similar to the deposition of silt in a fluvial river. In the Undara System, this is the best example discovered to date.
Photo: Dave Smith, C.C.C.

From the outset we recognised that it would not be feasible (initially anyway) to map all known caves. Hoping to discover how they had formed and whether there is any regular gradation in their transverse or longitudinal sections, caves were selected from three areas, namely: near Undara Crater, far from it and midway. Though we were unaware of it, Professor Macdonald of United States Geological Survey had observed and recorded lava tubes forming during eruptions on the Island of Hawaii. Features of the Undara tubes indicate that they formed in the same way — as depicted on page 6.

In the Undara caves we found no regular gradation, corresponding with distance from the Crater, in either their transverse or longitudinal sections.

At the suggestion of R. Greeley, one of NASA's lunar geologists, data were collected for the world's first longitudinal profile (shown **opposite**) through a crater of origin and representative caves of its lava tube system, by measuring slopes on a surface traverse corresponding with each cave traverse. Roof heights were obtained by measuring with helium-filled balloons. Roof thickness was then calculated graphically.

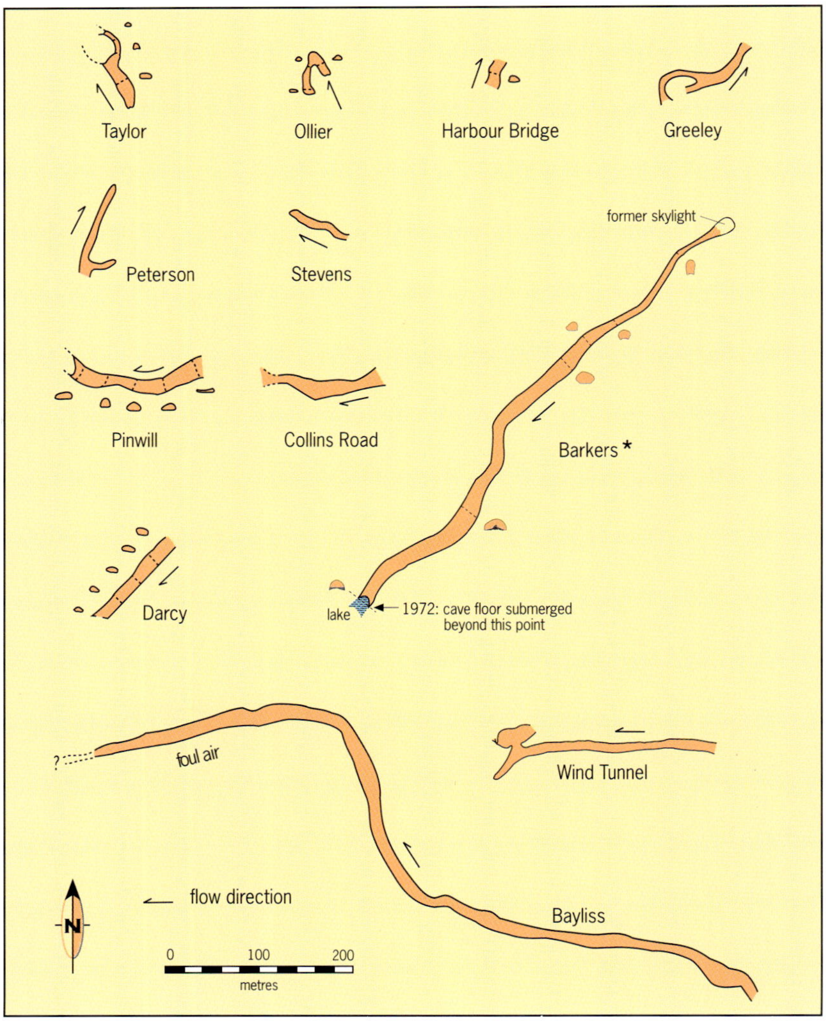

Plans and some cross sections of representative caves of the Undara Lava Tube System. Arrows indicate, flow direction.

*For detailed map and transverse profiles see page 79.

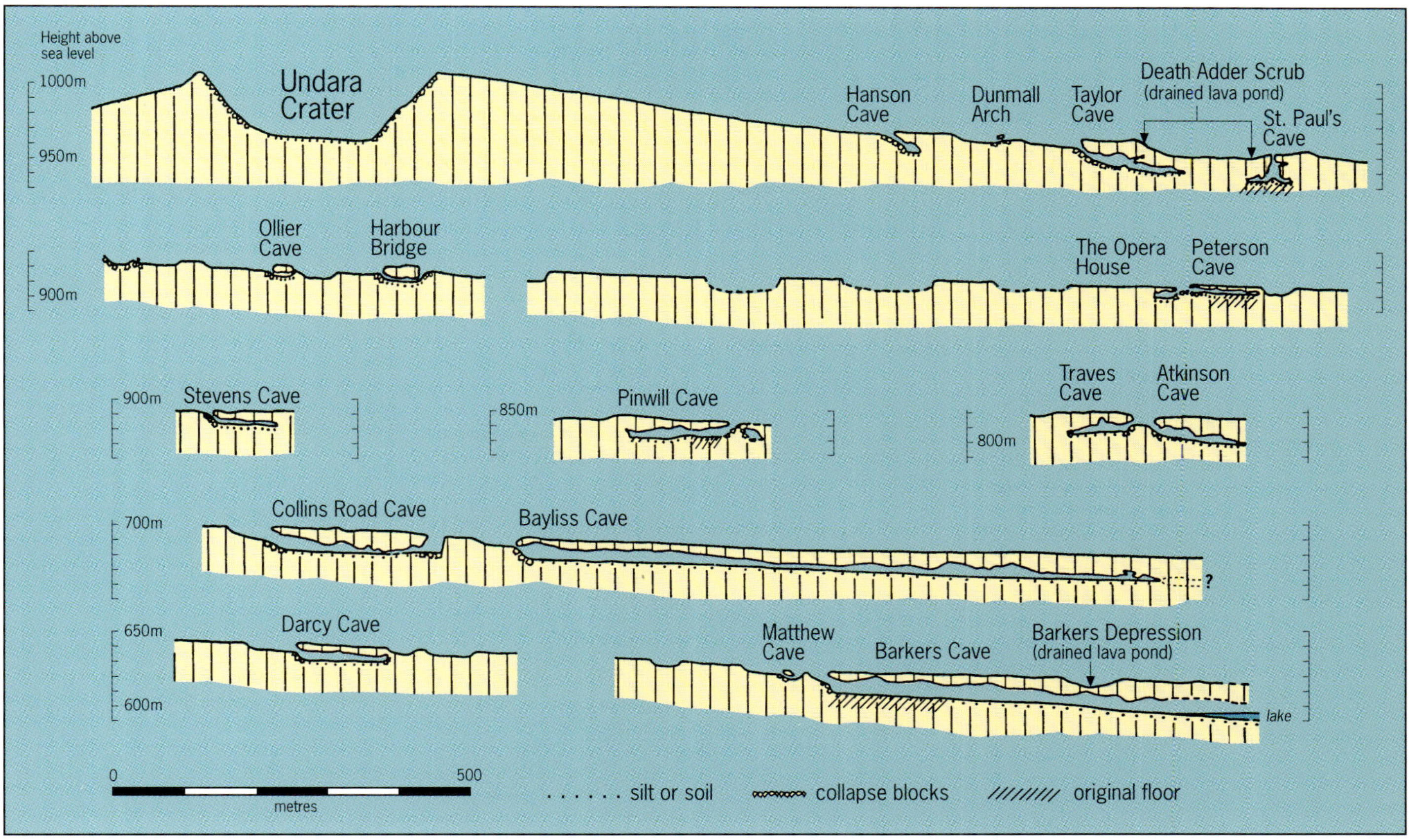

▲
Longitudinal section depicting Undara Crater and some caves up to 35 km distant from it, in the northwest flow. First published 1975.

Minor lava cascade. Under the thin crust on which the geologists are standing, lava drained back leaving us this small but perfectly preserved feature, the only one of its kind discovered to date at Undara. ▸
Photo: H.J.L. Lamont, J.C.U.N.Q.

Exploration in Bayliss Cave required oxygen masks because carbon dioxide percentage was dangerously high.
Photo: K. Bayliss, Queensland Fire Service.
▼

▲
Inflating strong balloons with helium gas for measuring cave heights.

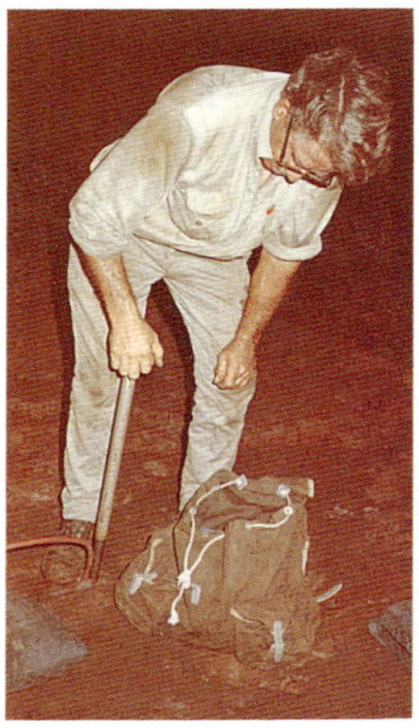

▲
Vernon collecting air in car tube in a cave where breathing became difficult. Analysis of a sample of this air by Queensland Health Department showed carbon dioxide percentage too high for exploration beyond this point, unless additional oxygen was available.
Photo: K. Bayliss, Queensland Fire Service.

B. Clark in Collins Road Cave measuring the height with a helium-filled balloon attached to a fishing reel by a graduated ribbon. ▸
Photo: M. Rimmer, R.S.S.

▲
Member of Chillagoe Caving Club exploration team, 1989, negotiating a very difficult crawlway — "The Duckunder".
Photo: P. Mansbridge, C.C.C.

1989, muddy but victorious!… Members of the Chillagoe Caving Club, Duncan Ray, Alex Lindsay, Daniel Donoghue and Phillip Mansbridge successfully explored more than 1.3 km from Bayliss Cave entrance. This established that Undara has the longest lava cave yet discovered in Australia. The floor was damp and very slippery and, for about half the distance, additional oxygen was needed. ▸
Photo: P. Mansbridge, C.C.C.

▴
Never were fresh air and daylight more welcome!
Photo: P. Mansbridge, C.C.C.

M. Rimmer places a steel post to mark the line of an underground survey. ▸
Photo: P. Mansbridge, C.C.C.

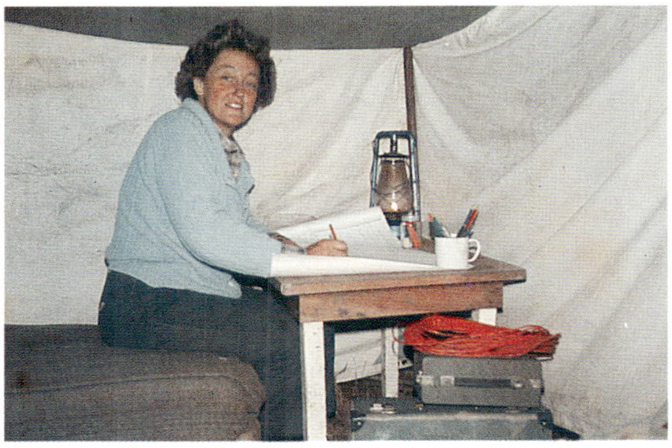

***Far right**: Plotting the day's data by the light of a hurricane lantern, 1973. Beside the table is a suitcase containing basic first aid materials and on top of it our "School of the Air" radio and aerial, a life-line in an emergency.*
Photo: B. Stanbridge, R.S.S.

▸ CAVE MEASUREMENTS

Research by us and others up to 1994 provided data for the following tabulation of measurements of caves discovered to date:

Cave	Register Number (A.S.F.)*	Length metres	Max. Width metres	Max. Height metres
Hanson	U1	40	12	3#
Dunmall Arch	U2,U3E	10	6	2
Taylor	U4	108	16.3	10.8
St. Paul's	U5	55	18.7	19.5
Sarah	U6	12	—	1
Peter	U7	13.8	9.9	3.8
Ollier	U8	49.4	10.4	3.2
Harbour Bridge	U9, U10E	35	14.3	5#
Greeley	U11, U12E	103	12.4	3.8
Frances	U13	6	12	4#
Opera House	U14	30	10	7.5
Peterson	U15	102 + 25	17.1	3.7
Stevens	U16	70.4	8.8	3
Pinwill	U17	150	21	8.9
Traves	U18	70	14	10.6
Atkinson	U20	101.2	28	6
Stephenson	U21	160	>25#	>10#
The Archway	U22	25	25	11
Ewamin	U23	100 + 50	22	>8#
Picnic I (down)	U24	360	17.6	15#
Picnic II (NE)	U25	40	12.7	14#
Dave I (up)	U26	50#	10#	8#
Dave II (down)	U27	70#	20#	—
Collins Road	U28	220	21.2	9.4
Bayliss (1973) additional (1988)	U30	>950 >400	25	11.5
Darcy	U31	99	16.3	6.3
Matthew	U32, U33	40	7#	3#
Barkers	U34	682*ƒ*	19.8	13.5
Raleigh I	U35	23	14.7	8.5#
Raleigh II	U36	29.8	17	10#
Lost World	U37	74.2	13.5	<9#

Cave	Register Number (A.S.F.)*	Length metres	Max. Width metres	Max. Height metres
Tween	U38	24	10.5	6.5#
Eptesicus	U39	42	23.5	6#
Inner Dome	U41	60	14	7.5#
Wind Tunnel	U42	226 + 40	32	8#
Short Little Arch	U43	15.8	7	2.1
Mikoshi	U44	46.6	14.4	11#
Misplaced Arch	U45	22	20	6#
Nasty	U46	120 + 10	15	8#
Fortune	U47	45.2	4	1.3
Temple of Doom	U48	49.5	5	4.5#
Fun	U49	33.2	9.8	1.25
Ding Bat	U50	50	13.5	7#
Hot Hole	U51	150	13.5	3.5#
Wishing Well	U52	>104	18	3.3#
Moth	U53	7.5	4	1.8
Sunset	U54	>30	4	4.8#
Wallaby's Hide-a-way	U55	27	9	4#
Expedition I	U56	30	12	3.5#
Expedition II	U57	33	20	4#
Arch (unnamed)	U58	7.9	10	2
Tom Tom	U59	34	9.5	1.8
Arch (unnamed)	U60	26	10	1.5
Komori	U61	80	9	2
Speaking Tube	U62	16.3	7.7	3#
Flat Ceiling	U63	80	15	3#
Branch	U64	10	10	5#
San	U65	25	10#	4#
Graham	U66	22	3#	3#
Upper Secret	U67	150#	—	—
Lower Secret	U68	70#	—	—
Total Length Surveyed:			**>6040.80 metres**	

Legend

\# estimated.

ƒ September 1994 — includes lake, 90m; depth of water at termination: 1m.

* The Australian Karst Register numbers generally refer to order of discovery, not distance from Undara Volcano.

▲
The Archway — at sunrise.
Photo: David B.J. Thomae.

▲
Skeleton Fork Fern, Psilotum nudum, *grows near some of the cave entrances at Undara. The fossil record shows that the species flourished during the Carboniferous Age when Australia was part of the super-continent, Gondwana, and has changed little in more than 300 million years.*
Photo: David .B.J. Thomae.

▲
The Archway — at noon.
Photo: C.E. Taylor.

Savannah Guide, Bruce Butler, pointing out a special feature. ►
Photo: by courtesy Q.T.T.C.

Facing page: Barkers Cave, 682 m long, is farther from Undara Crater than any other cave discovered to date. At this point two trains could pass and it is high enough to provide two levels of track. ▸
Photo: David B.J. Thomae.

◂ The flanged root of a rainforest fig near The Archway.
Photo: Craig Lamotte, Lamotte Editions.

The Archway — looking East.
Photo: Craig Lamotte, Lamotte Editions.
▾

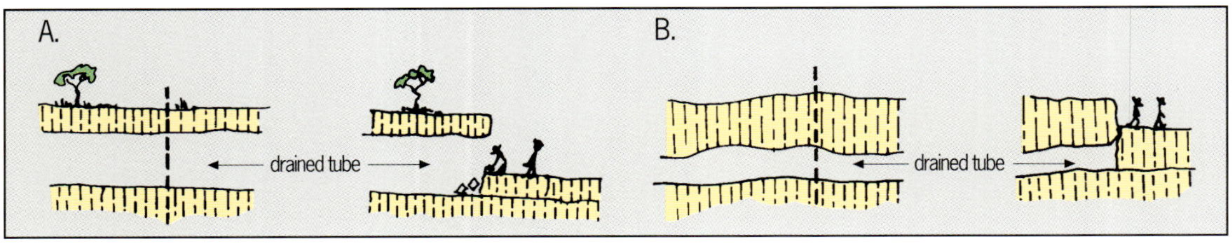

▴
There may be many more caves awaiting discovery! These diagrams show how entry to the tube can be blocked if the tube roof is thicker than the height of the drained tube. ***Diagram A*** *shows tube with thin roof.* ***Diagram B*** *shows tube with thick roof.*

◂ *Member of London-based Operation Raleigh abseiling into a cave in the course of their exploratory work in 1988.*
Photo: Richard Murray, The Australian Newspaper.

Rainforest fig framed in high cave entrance.
Photo: David B.J. Thomae.
▾

▲
Collins Road Cave. A Savannah Guide using a very powerful torch to illuminate a roof feature for a group of visitors.
Photo: Geoff Longford, Geo Productions.

Our record would not seem complete without this photograph of Barkers Cave. Taken in 1973, it has since appeared internationally. In 1992 it was chosen for the cover of the Proceedings for the 6th International Symposium on Vulcanospeleology held in Hawaii the previous year — a well deserved honour for former James Cook University photographer H.J.L. Lamont, A.R.P.S. (Lon.), A.A.I.P.P. This and his other outstanding cave photographs are a lasting tribute to his skill.

▲
Barkers Cave, 50m from its entrance. Note the gutter on the left and, on the distant wall, near-horizontal ridges, (lava level lines) evident almost to the roof. These represent periods when the lava river remained at a constant level. The lava tube is 13.5m high at this point — one of the greatest heights measured in the Undara Lava Tube System. At this location, granite hills on both sides of the flow restricted its width. A relationship is interpreted between this restriction of the flow's width and the height of the lava tube.
Photo: H.J.L.Lamont, J.C.U.N.Q.

▲
Accommodation at Undara today — vintage carriages from Queensland Rail, named after various nearby volcanoes, have been tastefully refurbished to provide comfortable modern sleeping and dining facilities at Undara Lava Lodge. Caravan and tent sites are also available.
Photo: C.E. Taylor.

▲
1990, three generations of the Collins family, Stan (standing) with son Gerry (right) and grandson Bram, who have exchanged their cattle grazing livelihood for management of the tourist guide service and accommodation at Undara Lava Lodge.
Photo: Margot Douglas.

UNDARA
EXPERIENCE

"...numerous volcanic foci... rise as conspicuous hills all over the plateau."

A. Gibb Maitland, 1891.

▲
A horizon of volcanoes — the Dry Season view, looking East from the granite bluff behind Undara Lava Lodge.
Photo: Mick Williams, T.C. & K.R.G.

Sunrise over One Hundred Mile Swamp during the heavy Wet Season of 1991, photographed from the granite bluff behind Undara Lava Lodge. The swamp is one of many locations in the area where drainage has been blocked by lava flows. It derived this name from the distance to the coastal town of Cardwell. The first telegraph line from Cardwell to the area was erected about 1870 and one old telegraph pole, complete with insulator, remains today. ▶
Photo: Mark Johnson, by courtesy Undara Experience.

◀ *The lava tubes, the associated depressions and adjacent open forest country are home to one of the most diverse collections of the kangaroo family (macropods) in Australia. Eleven species have been identified (see list, page 85).*

◂ *Water birds were not slow to find One Hundred Mile Swamp on one of the rare occasions when it filled, following the heavy Wet Season of 1991. Numbers of trees had their roots submerged for many weeks but few trees died.*
Photo: Mark Johnson, by courtesy Undara Experience.

▸THE ALIGNED RAINFOREST-FILLED DEPRESSIONS

That **the rainforest-filled depressions** at Undara are drained lava ponds was confirmed by Dr. D.W. Peterson, Scientist-in-Charge (1970–1975) at the United States Geological Survey's Volcano Observatory on the Island of Hawaii, when he visited Undara in 1976. He had seen lava ponds aligned with and adjacent to active lava flows in Hawaii and he agreed that small mosaics of distinctive, low porosity blocks represented areas of crust of a former lava pond. Such "mosaics" are rare, probably because drainage of most ponds was too fast to allow blocks of crust that were originally adjacent, to remain together after the ponds drained.

▲
***Island of Hawaii**, ca 1890. A lava lake retained by its own levée ring, within **Kilauea Caldera**. This photo gave us the clue to the formation of Undara's numerous aligned depressions with their dark green rainforest-type vegetation (see page 5), like the one near Barkers Knob, pictured below.* *Photo: by courtesy of the Bishop Museum, Honolulu.*

▲
Barkers Depression viewed from Barkers Knob. This oval depression with dark green vegetation formed when a lava pond drained. In this area, the flow was restricted between Barkers Knob and the hill, right middle distance.

Inset: Detail of inner wall to show a small "mosaic" of distinctive blocks of former pond crust.
Photos: Sarah Collins.

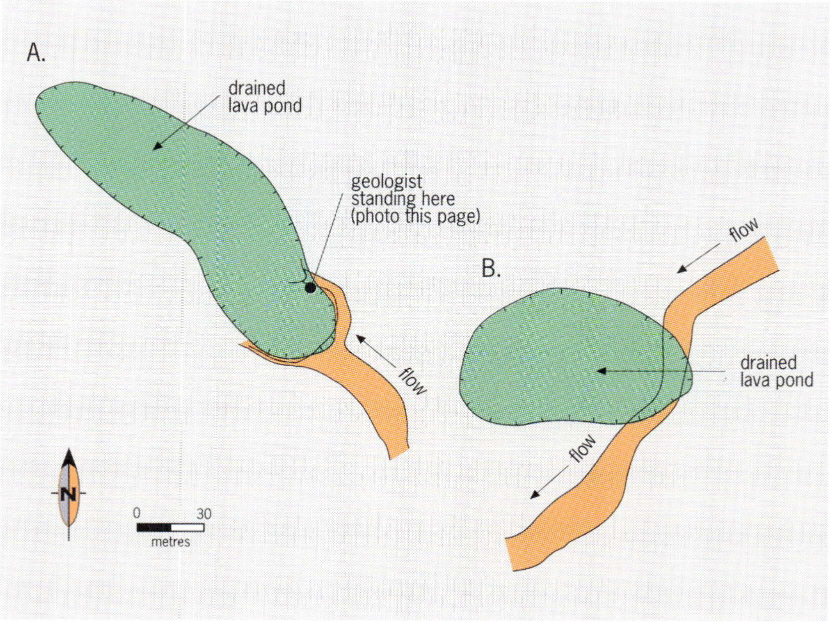

◂ *By superimposing maps of two of Undara's depressions on the surveys of two nearby caves, it was possible to confirm that the Undara depressions (and probably numerous similar features in North Queensland's basalt areas) had originated in the manner observed on the Island of Hawaii today. In diagram A. the former pond was in alignment with the tube (Taylor Cave). In diagram B. the pond was adjacent to the tube (Barkers Cave).*

Geologist Griffin near "The Pillar" at the termination of Taylor Cave. Note the prominent lava level lines on both walls and on "The Pillar". "•" on diagram A. above, indicates Griffin's position. *Photo: H.J.L. Lamont, J.C.U.N.Q.*
▾

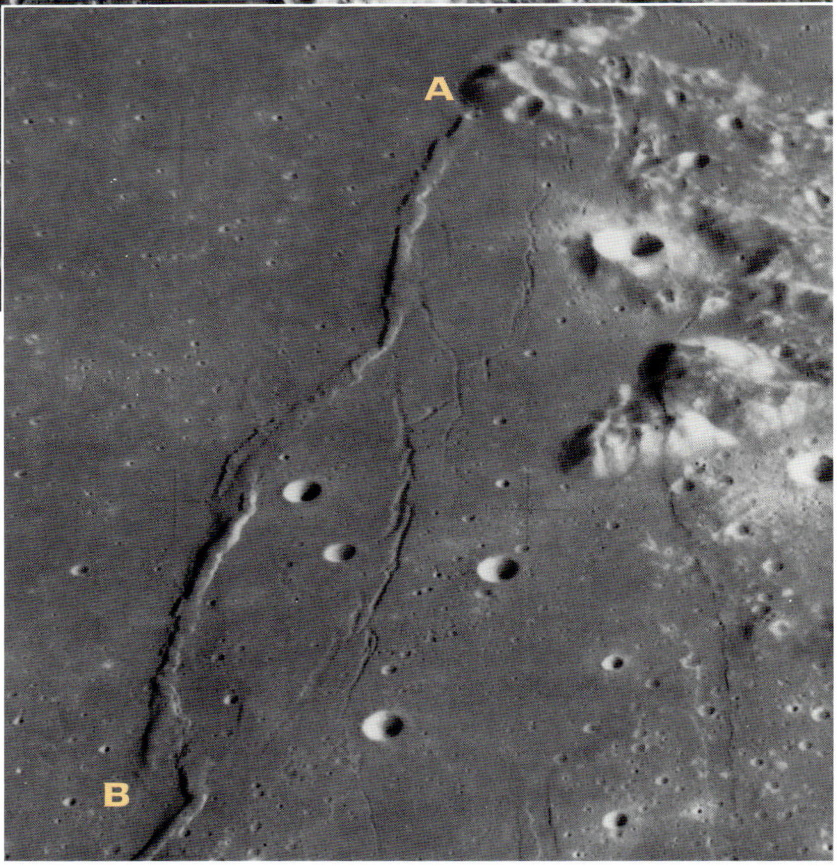

▲
***Area of the Moon** showing **Highlands and Basalt Plains** (darker areas) where **A–B–C** indicates a ridge. The numerous "pock" marks in the basalt plains are not volcanic craters but have been formed by the impact of objects from space and are known as impact craters.*

*Inset: Detail of section **A–B** ▸ of the basaltic ridge **A–B–C** marked on the photograph above.*

Photos: By courtesy NASA Frame No. AS17-0939.

▸THE MOON, VENUS AND THE WALL AT MOUNT SURPRISE

In **basaltic areas** of the **Moon**, NASA geologists had detected **ridges** (**A–B–C** in photograph, opposite) as well as channels like that on the lunar surface, shown in the photo on page 2. They had searched aerial photos of many of the Earth's basaltic areas for a similar ridge feature, but had found none.

In 1972, quite unaware of their search, the authors drew NASA's attention to the almost level ridge, 35 km long, known as **The Wall**, (below and pages viii, 50 to 52) Northwest of Mount Surprise, North Queensland. In a recent communication, NASA advises that, even now, no better Earth equivalent has been found.

Mount Surprise township, North Queensland. ▼

Recent analysis of radar data from the Magellan space probe to Planet **Venus** adds a new interest to the study of volcanic features on Earth. It is proposed that aligned depressions in volcanic areas on Venus may be connected by lava tubes and a sinuous ridge (page viii) has been detected there!

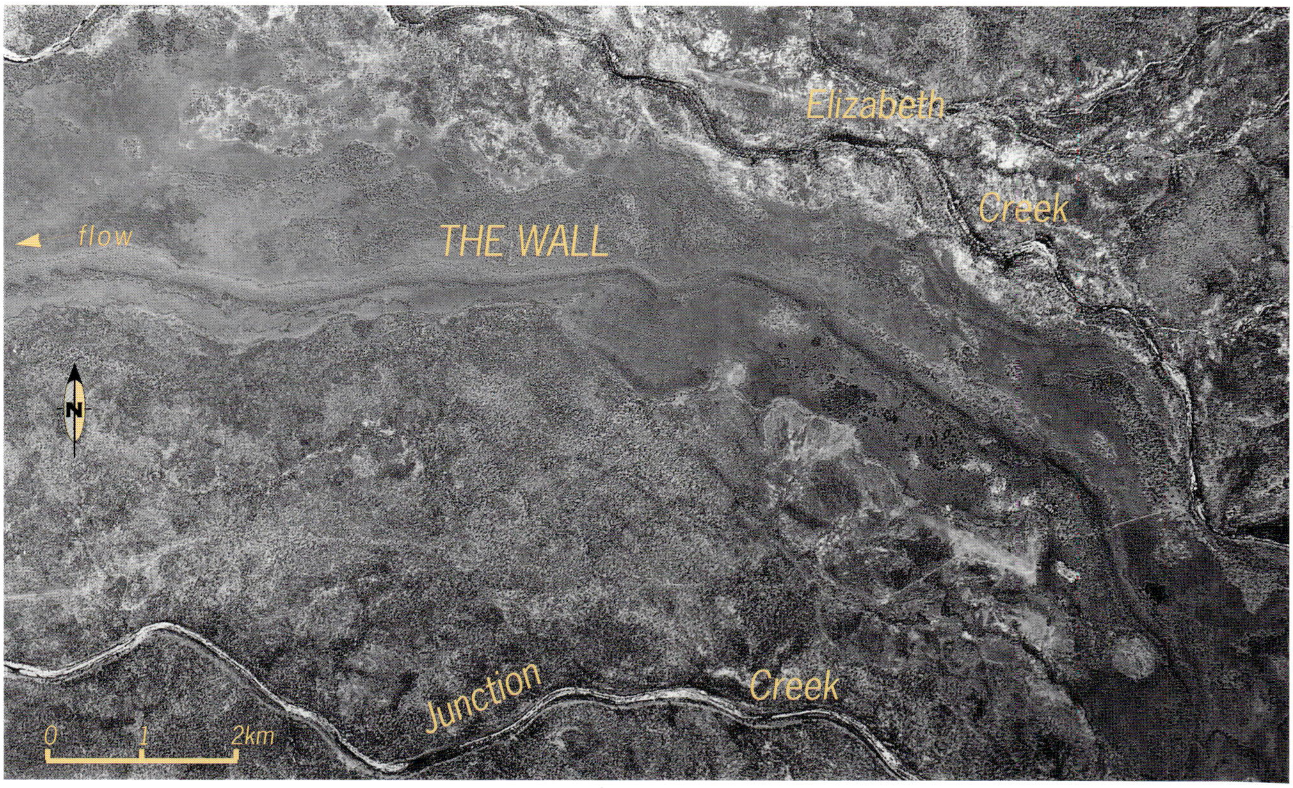

▲
*Aerial photo to show **The Wall's** location between two major creeks. Successful water bores close to The Wall encountered water-worn gravels under the basalt, confirming The Wall's location in a pre-basalt drainage channel.*
Photo: A.U.S.L.I.G. (see page 76).

▲ *A section of the 35 km basalt ridge, The Wall, near Mount Surprise, about 5 km East of its termination.*
Photo: Vernon Atkinson.

◄ *The western end of The Wall, viewed from low altitude. Polygonal basalt columns up to 2 m across flank a sunken median strip.*
Photo: Jon Edmonds.

The same section of The Wall, viewed from ground level. Arrows indicate basalt columns along the southern margin. Note the slight elevation below the third arrow.
Photo: Vernon Atkinson.
▼

◂ On top of The Wall, near its termination, polygonal basalt columns flank the southern margin.

Detail of one column. There is some elongation and alignment in flow direction of the gas holes, or vesicles. Notebook for scale is 17 cm.
▾

◂ Closer view of some columns.

DIAGRAMS TO SHOW POSSIBLE MODE OF FORMATION OF THE WALL,*
PARTICULARLY ITS WESTERN TERMINATION.

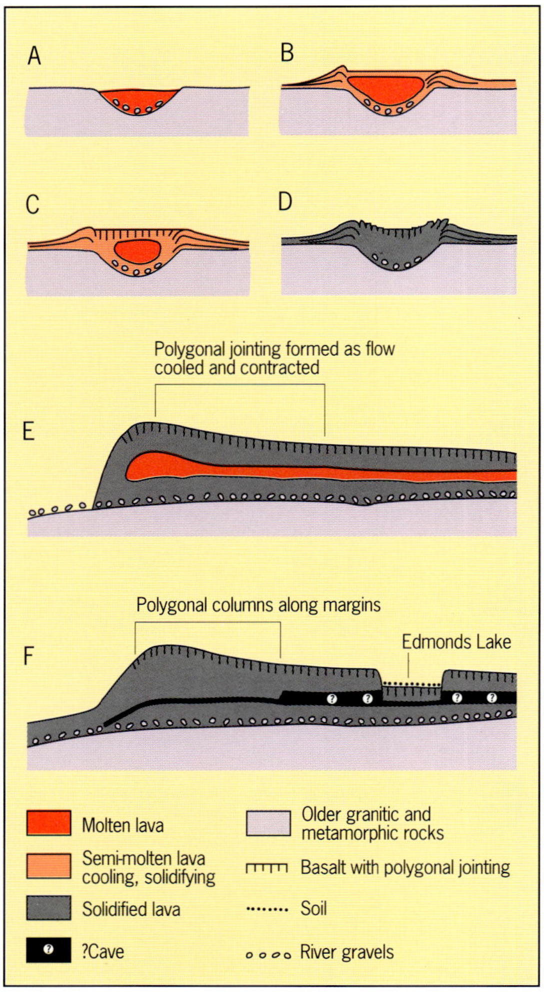

Transverse sections

A. Lava flowing in a valley...

B. ...built up levées along its main channel.

C. Lava hardened against the valley walls, at the base and top of the flow and in the levées but the central tube allowed lava to flow on. As the top of the flow cooled, it contracted and polygonal joints formed. The resulting columns show remarkable similarity in cross section measurements throughout the length of the flow. When the end of the flow hardened, the lava was dammed behind it.

D. A surge of lava broke through the lava dam, allowing still molten lava to escape. The channel roof sagged, leaving the prominent marginal polygonal columns.

Longitudinal Profiles

E. [corresponding to **C.** above]. The roof of the elevated channel cooled and polygonal joints formed. Further pressure of lava caused slight elevation at the termination.

F. Renewed activity from the vent broke the terminal dam and allowed the lava to flow on to the Einasleigh River. With the final withdrawal of supporting lava a central oval section near the termination collapsed to form Edmonds Lake, (see photo **below**).

* Geophysical research may solve this enigma.

*Edmonds Lake near termination of **The Wall** [See F. above]. This oval depression holds water for some months after the Wet Season.*

Photo: Vernon Atkinson.

The lunar basalt, shown below, is one of the basalt specimens in the 77 kg load brought back to Earth in 1971 by NASA's Apollo 15 Mission. This is the only **vesicular** basalt that was collected, although much of this type was sighted.

In all, the 12 astronauts who walked on the Moon collected 2,196 samples (382 kg) of lunar rocks, pebbles and dust from 6 locations. These have allowed scientists to put forward the theory that the Moon formed in a manner similar to the Earth. Of the total number of samples collected, 211 were basalt.

With no plans to return to the Moon, the lunar specimens that have not been sliced for scientific investigation in laboratories around the world, are guarded with the strictest security at NASA's Johnson Space Centre, Houston, Texas, U.S.A.

We pay tribute to the courage of those who went to the Moon and to those whose skills brought the most ambitious, the most challenging and the most complex of all space missions up to that date, to a successful conclusion...

"That's one small step for man, one giant leap for mankind."

Astronaut Neil Armstrong, as he became the first man to set foot on the Moon, July, 1969.

▲ ***Basalt** from the **Moon**. Despite its age of approximately 4,000 million years, this specimen could be mistaken for a fragment of newly erupted terrestrial lava. Its surface is not dulled by weathering because the Moon has no atmosphere. The numerous holes, or vesicles, formed when gases, dissolved in the magma, expanded as pressure decreased during the magma's ascent toward the Moon's surface. When the lava cooled and solidified the bubbles were trapped within it. (Actual size).*

Photo: by courtesy of NASA.

▲ ***Undara Basalt**. These blocks, on the inner slope of Undara crater, have many gas holes, or vesicles, giving them some resemblance to the lunar basalt pictured above. Zones of vesicular basalt are very common in basaltic lava flows on Earth, especially at the tops of flows. (Pocket knife for scale is 10 cm).*

Photo: Tom Atkinson.

LABORATORY STUDIES OF THE UNDARA BASALT

As **the McBride Volcanic Province** is so vast (over 5,500km²), Dr. Tim Griffin of James Cook University of North Queensland used aerial photos for preliminary delineation of the Undara and other lava flows in his Ph.D. study area. To confirm these delineations, he then collected over 400 specimens of basalt and associated rocks for laboratory examination.

By grinding slices of the rocks paper thin (photo at left), Griffin was able to follow the accepted practice of geologists and examine his specimens under a petrological microscope in ordinary light and also between crossed polars. To the naked eye, basalts are frustratingly similar but Griffin was able to detect microscopic differences that distinguished Undara specimens from those of other flows.

It is hard to believe that such uninteresting grey-black rocks have such beauty under the microscope. Few people ever have the opportunity to see this beauty, so the following photographs are included for your interest and enjoyment.

Following his microscope study of the McBride basalts, Griffin carried out geochemical analysis of selected specimens. Finally he and Dr. Ian McDougall of the Australian National University, using the potassium-argon method, determined the ages of some of the flows that had been delineated. They found that the Undara flows were approximately 190,000 years old, the third youngest in the Province.

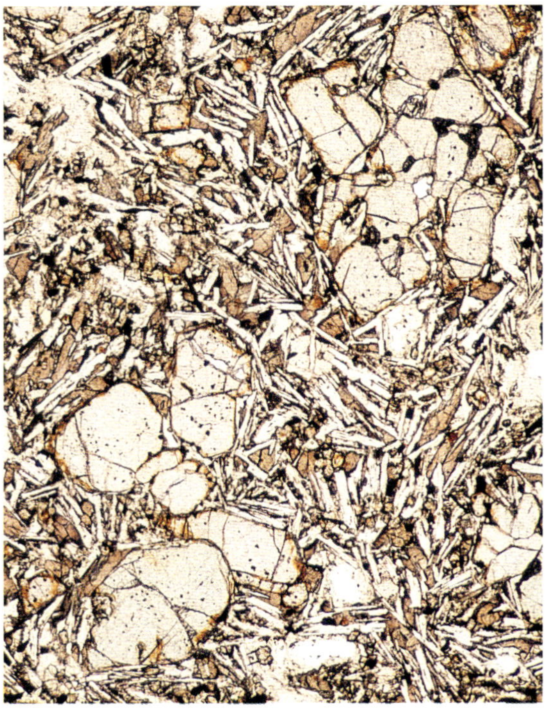

▲
Undara basalt as seen under a microscope. The larger crystals, called phenocrysts, are of olivine which crystallises before (at a higher temperature than) the mass of small grey-white feldspar and pinkish-beige pyroxene crystals. The colour of the pyroxene reflects its titanium content. ***Left:*** *Viewed in ordinary light.* ***Right:*** *Viewed between crossed polars. Magnification x 26.*

Photos: P.J. Stephenson, J.C.U.N.Q.

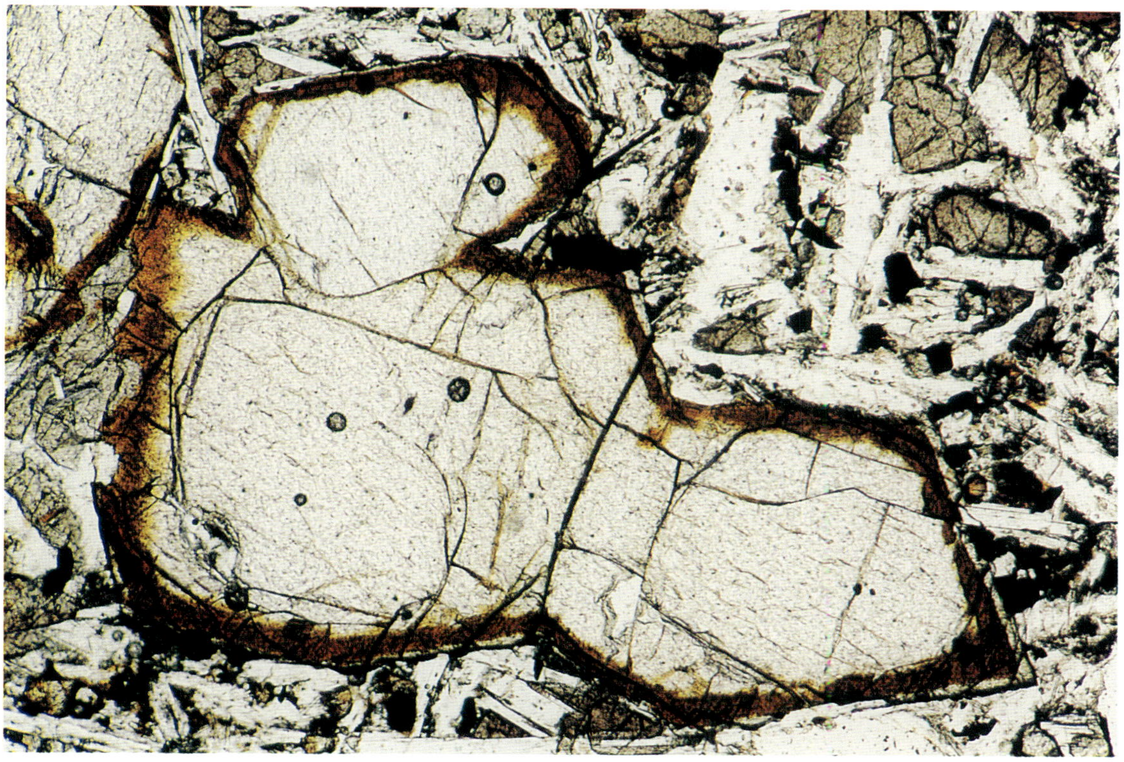

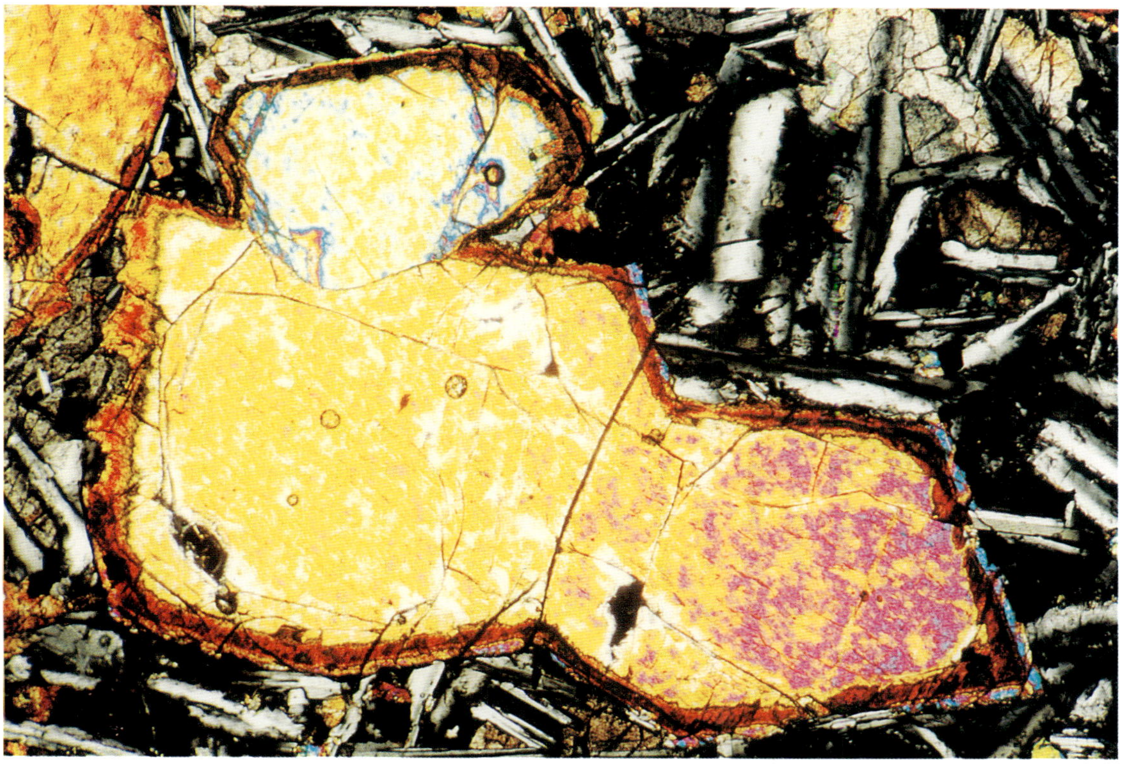

▲
A large crystal (phenocryst) of olivine, viewed in ordinary light [top] and between crossed polars [bottom]. Note the marginal zoning. The olivine in the inner band has been altered to opaque iddingsite. The narrow outer band is of olivine with a higher percentage of iron which gives rise to its beautiful colours in the lower photograph. Magnification x 70.
Photo: Anne Atkinson, J.C.U.N.Q.'s Leitz camera.

Acknowledgements and tributes

Although Aborigines would have witnessed the final spectacular eruption that occurred in the McBride Volcanic Province, no record of it has been discovered in the legends of their descendents.

A few artifacts have been found in the entrance to one cave, but no drawings or fires... and local Aborigines insist that their ancestors would have avoided all but the arches where daylight penetrated, unlike the American Indians, whose pictographs and ancient fires indicate the use of lava caves from prehistoric times.

The explorer Ludwig Leichhardt was the first person to record (1847) "young" volcanism when he travelled West through this area on his way from Moreton Bay to Port Essington. From his Journal, it would seem that he viewed the eastern and northern margins of the Undara lava field, although it was a younger flow from Kinrara Volcano, (see map, page 1) that he noted.

Ludwig Leichhardt
Photo: by courtesy John Oxley Library

In his Report of 1891 on the Upper Burdekin area, A. Gibb Maitland noted the 2,000 square mile (5,180km^2) extent of basaltic lavas on the "Burdekin-Einasleigh Plateau", (today termed McBride Volcanic Province), the numerous "volcanic foci" and the problems of travel, "especially with horses". He observed "tubular caverns" in the basalt and interpreted their mode of formation by the drainage of roofed lava channels. Later observations in Hawaii were to prove Maitland's interpretation correct, though it is still not accepted unanimously.

In their major work on the Geology and Palæontology of Queensland & New Guinea, 1892, R.L. Jack and R. Etheridge appear to have studied volcanic provinces to the North and South of the McBride Province. They indicate the McBride Province on their map but their description of it is quoted from Maitland's Report of 1891.

In 1956 geologist, C. R. Twidale, observed the alignment of the depressions North and Northwest of Undara Crater but incorrectly interpreted them as evidence of a fissure eruption.

It was early in the 1960s, in the preparation of the geological map of the area, that vulcanologist J. (John) G. Best, from the Bureau of Mineral Resources (B.M.R.), correctly interpreted the alignment of the depressions as indicative of what he termed "collapsed lava tunnels". Time constraints did not allow the group to explore the features but with great perception — or was it intuition? — Best named the volcano **Undara**, Aboriginal for "long way". Did he sense the magnitude of the flow which has now proved to be the world's longest from a single volcano?

John Best

The term "collapsed lava tunnel" on the B.M.R. 1962 map intrigued bush-walker/ engineer Norman Traves who, in 1966, was supervising construction of the new Gulf Developmental Road. Traves went in to see the "tunnels" and was amazed by their magnitude. Recognising them as worthy of National Park status, he recommended this to the Minister for Local Government and Conservation through the National Parks Association of Queensland. In 1967, field officer Peter Stanton confirmed their unique character.

Norm Traves

The Collins family, who have grazed cattle in the area since early in the 1860s, and their stockmen have always known that caves existed, but they were not widely known until 1985 when Richardson and Newman acclaimed them as one of the seven wonders of Queensland in their book "Australia's Natural Wonders". The ensuing publicity brought so many tourists that, to help visitors, the Collins family is now providing accommodation and a guide service at their Undara Lava Lodge.

We owe many of the facts and figures of Undara to geologist, Tim Griffin, whose project for his Ph.D. degree was a five year study (1970 to 1975) of the 5,550 km^2 McBride Volcanic Province (see map, page 1). This was suggested and supervised by Associate Professor Jon Stephenson, James Cook University of North Queensland. Local people never ceased to wonder that, time after time, Griffin from the city and often on a motor bike, returned safely from his lonely sojourns in the area where he collected more than 400 basalt samples for scientific investigation.

Peter Stanton

The authors' research (1972–1974) at Undara stemmed from an urgent request from Dr. W. Halliday, National Speleological Society (U.S.A.) to Dr. N.C. Stevens, University of Queensland, for a paper for the first International Symposium on Vulcanospeleology, 1972. Unable to visit the area immediately, Stevens passed the request on to Anne and we are grateful for the continuing interest of Halliday, Stevens and all others who have assisted us.

As mentioned earlier, we selected three sections of the lava tube system, namely: near the Crater, far from it and midway, hoping to discover how the tubes formed. Field methods were based on Anne's matriculation studies at "Redlands", Sydney, under Miss Dorothy Taylor, an inspiring teacher.

Our task at Undara would perhaps never have been completed but for the interest and assistance of our own family and friends, R.Dunmall, T.Griffin, the Pinwill, Edmonds and Collins families, other local residents, Aboriginal assistants, Darcy Day and others — a veritable army of volunteers, too many to be listed here, few of whom returned for a second bush weekend! The support of all the above people, of Vera Kay, Frances Johnson, Audrey Durham, Keith McLean, Aaron Fallows and D.Stephens (who typed the first manuscript), and Greg Atkinson (who converted all field measurements to metric, for drafting) is gratefully acknowledged, as is the supervision of the project by Associate Professor Jon Stephenson, J.C.U.N.Q., after Professor Willard Lacy accepted Anne into his Geology Department and later convinced her that this book should be compiled.

Pat Comben

The Honourable Pat Comben, while Minister for Environment and Heritage (1990–1993), saw the need for protection of the lava tube system and proposed the gazettal of **The Undara Volcanic National Park**. This was officially opened in 1992 by the Premier of Queensland, The Honourable Wayne Goss.

Members of the Chillagoe Caving Club of North Queensland have made valuable contributions, discovering and mapping caves and assisting biologists.

Advice from Dr. F. L. (Lyn) Sutherland of the Australian Museum led to improvements in this manuscript.

Journalist Vivienne Skinner advised on a more popular style of presentation. This and the expertise of Designers Paul Ramsden and Janice Watson of The Queensland Museum are gratefully acknowledged and have made this book far more readable.

While keeping long term preservation as the foremost goal, the Department of Environment and Heritage, Gerry and Bram Collins and the Savannah Guides have made an invaluable contribution. They have enabled thousands of visitors to view and appreciate the lava caves and environs.

Richard Campbell, Producer, Australian Broadcasting Corporation's Natural History Unit and his team have produced a comprehensive television documentary. It brings to viewers here and overseas, the amazing story of a volcano that must in time become one of Australia's best known geological features.

To indicate the international nature of the Undara investigations, the following eight points should be recorded:

- Much of the story might still be conjecture but for the interest and co-operation of Dr. Donald (Don) W. Peterson. From 1970 to 1975 Peterson was Scientist-in-Charge at United States Geological Survey's Volcano Observatory on the Island of Hawaii. With his knowledge of recent and active lava flows he confirmed or refuted, through correspondence with the authors, various hypotheses about Undara's features. Peterson eventually visited Undara with the authors in 1976 and was able to confirm the accuracy of his long distance assessments.
- Professor Ronald Greeley, Geology Department, Arizona State University, U.S.A., former NASA Research Scientist, added an exciting facet to the research at Undara in 1972, when he declared The Wall near Mt. Surprise, North Queensland, to be Earth's best analogue to the smaller basaltic ridges on the Moon. We have appreciated his continued interest.
- Christina Heliker, a geologist with the United States Geological Survey's Hawaiian Volcano Observatory, has been monitoring the current eruption for ten years. We are deeply indebted to her for giving us, in 1991, our first glimpse of lava flowing in a lava tube and, more recently, invaluable advice on the Hawaiian section of this book.
- Advice from Professor R. Hazlett, Pomona University, Claremont, California U.S.A. and permission to use his diagram in Appendix 1 are gratefully acknowledged.
- Staff of Hawaii Volcanoes National Park assisted us on each of our visits to the Island of Hawaii and Dan Taylor advised on aspects of the Hawaiian section of this book.
- We are deeply indebted to the United States Geological Survey for permission to include in this book their outstanding colour images of active lava and to Taeko Jane Takahashi for selecting them from the archives at Hawaii Volcano Observatory.

- Over the past ten years, Dr. Nicholas Sullivan of The Explorers Club, New York, has obtained funds for research at Undara by cave biologists, Drs. Frank Howarth and Fred Stone, from U.S.A., and Drs. Hannelore Hoch and Mannfred Asche from Germany. Their work has shown that one cave at Undara can, at present, be claimed as the world's most biologically significant.
- In 1988 young members of the London-based Operation Raleigh and leader Brian Furniss, under the skilled guidance of Queensland National Parks Field Officer, Michael (Mick) Godwin, investigated three previously unexplored sections of the tube system, discovering many new caves, including the first caves in Undara's north flow and adding much to the biological records.

Finally we would mention that, just as the field work at Undara could never have been done by a geologist single-handed, neither could this book have been compiled without the interest and skill of the 61 photographers, artists and draftsmen who have enabled us to tell the story of Undara in pictures. We trust that our book will aid understanding and appreciation of this remarkable geological wonder. Our earnest hope...

that UNDARA may be for all people, for all time.

▲
Sunset at Barkers Knob in the Undara lava field. This is one of the many granite hills that were encircled by lava from the Undara Volcano.
Photo: M. Williams, T.C. & K. R. G.

▸BIBLIOGRAPHY (UNDARA)

for full list of references see Atkinson, F. Anne, 1992

▲
Part of the northern margin of the Undara lava field. Here it forms the southern bank of the Lynd River.
Photo: Vernon Atkinson.

ATKINSON, F. Anne, 1988a: The Remarkable Undara Lava Tube System — A Geologist's View, *in Pearson, L.M. (Editor) Proceedings of the 17th Biennial Congress, "Tropicon"*, Aust. Speleo. Fed., pages 39–56.

ATKINSON, F. Anne, 1988b: Vulcanospeleology — Extra Terrestrial Applications and the Controversy: Mode of Formation of Lava Tubes *in Pearson, L.M. (Editor) Proceedings of the 17th Biennial Congress, "Tropicon"*, Aust. Speleo. Fed., pages 57–63.

ATKINSON, F. Anne, 1990a: The Remarkable Undara Lava Tube System, North Queensland. Points of interest as a background to its Unique Biology. *Jour. Tasmanian Cave and Karst Research Group*, pages 7–26. *Distributed as: "The Lava Tubes of the Undara Volcano, Far North Queensland, Australia".*

ATKINSON, F. Anne, 1990b: The Undara Lava Tube System and its Caves. Abridged from Atkinson 1988a and b and updated. *Helictite 28(1)*, pages 3–14.

ATKINSON, F. Anne, 1992: The Undara Lava Tubes, North Queensland, Australia — Updated data and Notes on possible Lunar Analogue *in Rea, G. Thomas, (Editor) in Proceedings of the 6th International Symposium on Vulcanospeleology, Hawaii, August 1991*, pages 95–120.

ATKINSON, F. Anne, 1993: Geological Features of the Undara Lava Tube System, North Queensland — with particular emphasis on the features in the caves and environs to be used for tourist visitation from 1993. *Report for the Department of the Environment and Heritage (unpub).*, 112 pages (including 15 maps).

ATKINSON, F. Anne, 1995: The Undara Lava Tube System, North Queensland, Australia: Updated Data and Notes on Mode of Formation (Abridged from Atkinson 1991) *in Baddeley, G. (Editor) Precedings of the 20th Biennial Conference, "Vulcon", Aust. Speleo. Fed.*, pages 1–14.

ATKINSON, F. Anne, Griffin, T. J. and P.J. Stephenson, 1975: A Major Lava Tube System from the Undara Volcano, North Queensland. *Bulletin Volcanologique*, 39(2), pages 1-28.

GODWIN, M.D., 1993: Undara and Associated Lava fields of McBride Plateau, a speleological field guide. *Chillagoe Caving Club Inc.*, 360 pages.

GRIFFIN, T.J., 1977: The Geology, Mineralogy and Geochemistry of the McBride Basaltic Province, Northern Queensland. Ph.D. thesis *J.C.U.N.Q. (unpub).*, 225 pages.

LARSON, C.V., 1992: Nomenclature of Lava Tube Features in *Rea, G. Thomas, (Editor) in Proceedings of the 6th International Symposium on Vulcanospeleology, Hawaii, August 1991*, pages 231–248.

STEVENS, N.C. and F. Anne Atkinson (1975): The Undara Lava Tubes, North Queensland, Australia. In W.R. Halliday, (Editor) *Proceedings of the International Symposium on Vulcanospeleology and its Extraterrestrial Applications.* A special session of the 29th Annual Convention of the National Speleological Society, White Salmon, Washington, U.S.A., August, 1972.

STEPHENSON, P.J., Griffin, T.J. and F.L. Sutherland, 1980: Cainozoic Volcanism in Northeastern Australia in *R.A. Henderson and P.J. Stephenson (Editors) Geology and Geophysics in Northeastern Australia*, pages 349–374.

***Facing page**: We will always envy the visitors who had this view of Puu Oo — but they considered it rather too close for comfort!* ▸
Photo: Noël Black, Hawaii.

SOME FACTS AND PHOTOS FROM

THE ISLAND OF HAWAII

AN AID TO UNDERSTANDING UNDARA'S FEATURES

▸THE ISLAND OF HAWAII

You well may ask: "Why the Island of Hawaii in this book?" The reason is this: As we have no active volcanoes in Australia, we have to look overseas to see freshly erupted, or actively flowing lava… and it is to the Island of Hawaii that we look, for it has long been regarded as the best place on Earth to study volcanoes because eruptions there are very rarely dangerous and the climate is pleasant.

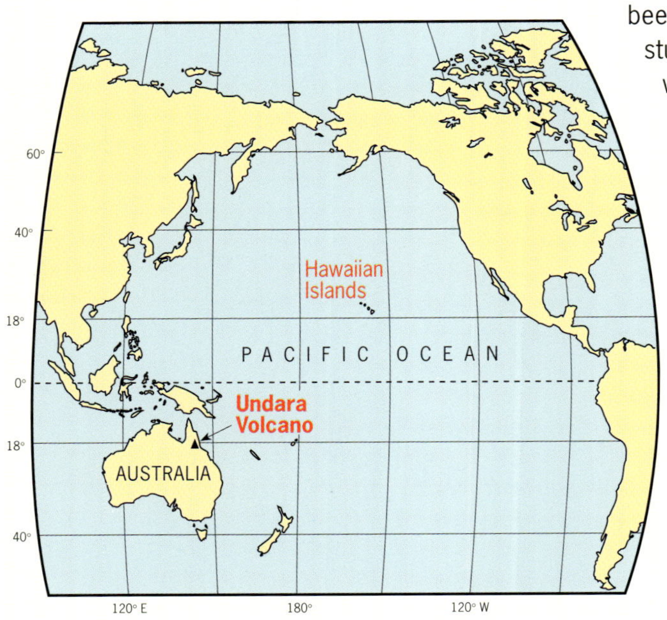

The authors have had the good fortune to see active lava on each of their four visits to the Island of Hawaii. The photographs on the following pages are included to give some idea of how the lava may have looked at Undara 190,000 years ago. Unfortunately no photo or description can do justice to the incredible sounds of eruption — the swirl of a river in flood, the crash of ocean waves, the plop of a porridge pot, — to mention but a few.

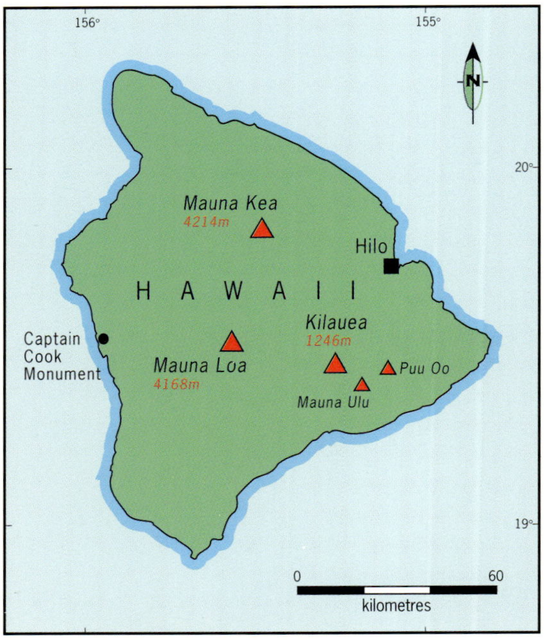

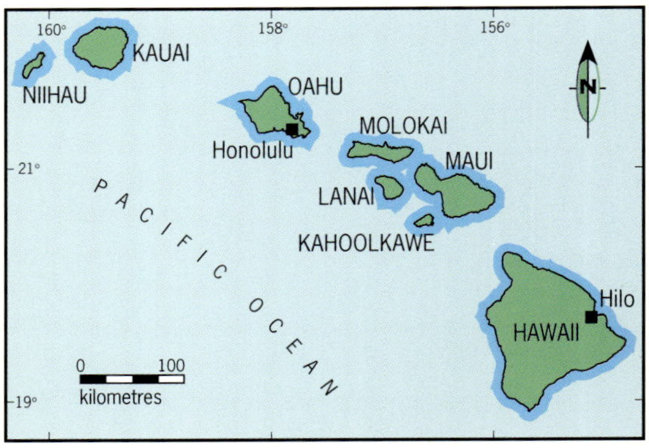

▲ *The southern islands of the Hawaiian group. Built from ocean depths by volcanic activity, these islands span more than 1500km, like jewels across the mid-Pacific. Their age increases progressively to the Northwest.*

◂ *Island of Hawaii, the southernmost member of the group. It is the youngest and most active island, having emerged from the sea only 800,000 years ago — today, geologically speaking!*

***Facing page**: 1985, Puu Oo vent, Kilauea East Rift Zone. This spectacular fountain is over 350m high. Note how the surface lava in the channels is darkening as its temperature drops on exposure to air. The black spots to the right of the fountain are fragments of frothy lava called scoriae. As the lava is hurled skyward, some is drawn out into fine threads and solidifies. These fine strands may drift some distance from the vent before settling in mats that look like human hair (see Glossary — Pele's Hair). Shining in the "hair" are tiny black droplets of lava — Pele's Tears.* ▸

Photo: Lee Allen Thomas, Hawaii.

▸LAVA TUBES ON THE ISLAND OF HAWAII

On the Island of Hawaii one lava tube is over 25km long. Within Hawaii Volcanoes National Park, however, tourists are encouraged to visit only one cave, depicted **left** and **bottom** — the Thurston Lava Tube which was discovered in 1912. Access to it has now been made easy and a section of the tube is permanently illuminated. Sad to relate, most of its fragile lava structures have been removed as souvenirs.

▲
Entrance to a cave near the trail to the summit of Mauna Loa.

◂ *Thurston Lava Tube Cave viewed from inside. Note the deep entrance profile.*

Interior of the Thurston Lava Tube Cave. Droplets of water catch the light to give an almost fairy-like appearance.
Photos (Thurston): Peter French, Hawaii.
▼

FLOW TYPES, THEIR TEMPERATURE AND SPEED OF TRAVEL

Although chemically identical, two distinct types of lava flow are recognised: **"pahoehoe"** with a smooth, billowy surface and **"aa"** with a blocky, jagged, clinkery surface. These Polynesian names were first used in Hawaii and are now accepted worldwide.

After an initial fountaining stage, lasting less than 24 hours, most of the flows in Hawaii emerge from vents as the fluid pahoehoe type of lava, often changing to the more viscous aa type as they move down-slope. This change results from the loss of dissolved gases and a degree of crystallisation which means that a flow can never revert back from aa to pahoehoe.

The **temperature** in the interior of active lava flows in Hawaii is almost 1200° centigrade and it drops only a few degrees even after travelling many kilometres from the vent. The **speed** of both pahoehoe and aa flows is determined by the rate and volume of the eruption, and the slope.

From the vents, **flows** are fed from main channels usually less than 15m wide and, on steep slopes, pahoehoe lava in these channels can travel at speeds up to 56km per hour. However, main channels branch to form **flow fronts** that may be as wide as 1km. These wide fronts advance at very much slower rates — at 3km per hour, or less.

Lava tubes form almost exclusively in the fluid pahoehoe type of lava so, in this book, the speed of pahoehoe lava flowing in lava tubes is of special interest. Active lava tubes were first studied in Hawaii during the eruptions of Mauna Ulu, 1970–1971, when tubes up to 12km long formed. Lava flowed through them at rates of 1km to 6km per hour. **During the current eruption, the average speed recorded of the pahoehoe lava, travelling in tubes from Kupaianaha vent, has been 3.5km per hour.**

A rough "aa" flow between two "pahoehoe" flows on Mauna Loa's north flank. The white buildings in the middle distance (altitude 3,383m) house the National Oceanographic and Atmospheric Administration Observatory. The volcano on the horizon is Mauna Kea, 4,214m.
Photo: Anne Atkinson.

▸FLOW TYPES — "PAHOEHOE"

▲
This fluid "pahoehoe" flow front has ignited the light vegetation in its path. The characteristic grey-black glistening surface of the lava shows some areas of ropy structure ***[A]***, ***[B]*** *and* ***[C]***. *This forms when lava continues to move below a thin, flexible surface crust.*
Photo: J.D. Griggs, U.S. Geological Survey.

◂ *Geologist measuring the temperature of molten lava. He finds it is about 1160° C.*
Photo: D.N. Little, U.S. Geological Survey.

Opposite: A "pahoehoe" flow advancing through an open forest gives us a glimpse of how Mount Surprise country might appear if volcanic activity resumed. ▸
Photo: J.D. Griggs, U.S. Geological Survey.

▲
Mauna Ulu lava lake, 1972. As the vent grew, the level of this lake of "pahoehoe" lava rose and fell and poured from one level to another forming a lava cascade, or lavafall. At Undara, a minor lavafall is envisaged near the entrance to Barkers Cave.
Photo: Bryan Harry, U.S. National Park Service.

▲
1991, August. At the end of a long walk on newly erupted lava, Vernon using the heat from a new "pahoehoe" toe to warm his hands. Despite the wind and the rain, we watched spellbound to see, at last, "pahoehoe" toes bud and ropy lava form — unforgettable sights!
Photo: Anne Atkinson

▲
Ropy lava. As underlying molten lava continues to move beneath the stiffening, but still-plastic crust, the surface of a "pahoehoe" lava flow sometimes becomes twisted into forms resembling coiled ropes.
Photo: J.D. Griggs, Hawaii

▲
Expanse of freshly erupted "pahoehoe" lava to show how much of the Undara lava field would have appeared straight after eruptions. A roofed channel is running diagonally across the foreground from the left corner.
Photo: Vernon Atkinson.

But life goes on... within a few days of eruption, while the lava is still quite warm, the sound of crickets can be heard at night... within months, vegetation appears on the lava flows if there is sufficient rain. Here, hardy ferns have germinated on one of the earlier flows from Kupaianaha. Looking East, toward the sea, trees on the margin of the forest in the middle distance have died from the intense heat of the flow. In the far distance a plume of steam marks a location where lava, travelling through lava tubes from Kupaianaha vent, was flowing into the sea (August, 1991).
Photo: Vernon Atkinson.
▼

◂ *July 19th 1974, sunset. Lava, erupting in a spectular line of fountains up to 30 m high, coalesced and flowed toward the photographer. View across the summit caldera of Kilauea Volcano on the first day of an eruption, which lasted only 3 days.*

Photo: Sandy Colvin, Hawaii.

July 20th 1974, lava flowing in channels from a small cone at the southern end of the fissure on the floor of the summit caldera of Kilauea Volcano. Note the slabs of lava solidifying even as they are moving along the channels and the red streaks of molten lava in the sloping walls of the channels, an indication of their semi-solid state. ***[X]****,* ***lower left, facing page****, indicates where Anne is standing in photo, top right.*
Photo: Anne Atkinson.
▼

▲
July 20th 1974, at the eruption of Kilauea Volcano, Anne after 36 hours with no sleep, and looking like it!
Dr. R.I. Tilling, a geologist at the United States Geological Survey's Hawaii Volcano Observatory, had taken a party of three to the point marked [X], left foreground, ***facing page****, to collect fresh lava specimens for his study of chemical variations during eruption.*
When they smelled somebody's boots were burning they did not wait to see whose!
Photo: R.I. Tilling, U.S. Geological Survey.

▲
July 21st, 1974. The same scene, just 24 hours later. The channel roofs had solidified and blackened. To our utter amazement, and disappointment, the eruption was nearly over. On the near slope a National Park ranger can be seen talking to a Polynesian lady.
Photo: Vernon Atkinson.

▲
August 1991, with Christina Heliker, U.S. Geological Survey geologist at a "sky-light" — a hole in the channel roof above a lava tube carrying lava from Kupaianaha vent, East Rift Zone of Kilauea Volcano. Heliker, working at the Survey's Hawaii Volcanoes Observatory, has been monitoring this eruption since 1984. The vent on the horizon is Puu Oo. Kupaianaha vent is out of view to the right.
Photo: Vernon Atkinson.

◂ *August 1991, closer view of the "skylight" pictured above.*
"Skylights" have to be approached from the windward side and with great caution — the heat is so intense. In this tube the temperature of the lava is about 1160°C.
Photo: Vernon Atkinson.

▲ *The compact centre of an "aa" flow exposed in a road cutting on Mauna Loa.*

Aa lava is far more viscous than pahoehoe lava so it may seem surprising that the fastest movement of a flow front recorded in Hawaii was that of a 1950 aa flow from Mauna Loa when a speed of over 9.7 km per hour was maintained for six hours.

The movement of an aa flow can be likened to that of the tracks of a giant tractor — pressure from behind forces the jumbled, clinkery upper surface of the flow under the pasty interior.

Typical active "aa" flow with jumbled clinkery surface and dense interior. ▸
Photo: J.B. Stokes, U.S. Geological Survey.

The viscosity of aa flows rarely allows lava tubes to form in them. However, following the 1973 eruption on the Island of Heimaey, to the surprise of Icelandic geologists, the authors discovered a lava tube in an aa flow from the Eldfell Volcano.

1973. Lava flow from Mauna Ulu descending the south flank of Kilauea. At an earlier stage, the massive flow had changed to rough, black "aa" lava. The newer lava that continued to erupt formed a channel down the centre of the "aa" flow and shows as a brilliant molten stream of "pahoehoe" lava. The trees in the top right section of the photo have been stripped of their leaves by the falling scoriaceous material which surrounds them. Photo: D.W. Reeser, U.S. National Parks Service.
▼

▲
Lava tree casts — Dr. D.W. Peterson, Scientist-in-Charge (1970–1975) at the U.S. Geological Survey's Hawaii Volcano Observatory, explaining how these formed when a "pahoehoe" flow engulfed the trees that were standing here and moulded to their shape, leaving these casts when the flow level dropped.
To date, cylindrical cavities, possibly of this origin, have been found at only two locations in the Undara lava field. Understandably, no vertical casts comparable to the above have withstood the 190,000 years of weathering.
Photo: Vernon Atkinson.

▸ HAWAIIAN VOLCANOES — Suggested additional reading

DECKER, Robert W. and Barbara Decker, 1991 (4th edition revised): Volcano Watching. *Hawaii: Hawaii Natural History Association in co-operation with the U.S. National Park Service, U.S. Department of the Interior*, 82 pages.

KAYE, Glen, 1976: Hawaii Volcanoes — The Story Behind the Scenery. *Las Vegas, Nevada: KC Publications, Inc.*, 48 pages.

MACDONALD, Gordon A., Agatin Abbott and Frank Peterson, 1983: Volcanoes in the Sea — the Geology of Hawaii. *Honolulu: University of Hawaii Press*, 518 pages.

MACDONALD, Gordon A., and Douglass Hubbard, revised 1989: Volcanoes of the National Parks of Hawaii. *Hawaii: Hawaii Natural History Association*, 66 pages.

WEISEL, Dorian and Christina Heliker, 1990: Kilauea — The Newest Land on Earth. *Honolulu: Bishop Museum Press*, 76 pages.

WRIGHT, T.L., Takahashi, Taeko Jane & J.D. Griggs, 1992: Hawaii Volcano Watch, A Pictorial History, 1779–1991. *University of Hawaii Press*, 162 pages.

‣ GLOSSARY

From Glossary of Geology, 1979, Macdonald and Hubbard, 1989, Larson, 1992 and other sources.

Aa Lava with rough clinkery surface. (Pronounced ah ah, this Polynesian word is now accepted worldwide).
Origin: allegedly from the cry of pain when crossing this type of lava bare-footed.

Ash Sand or dust size material ejected from a volcano.

Basalt A dark-coloured igneous rock, rich in iron and magnesium and comparatively poor in silcon; the common lava of the continents, the ocean floors and Hawaii.
Origin: believed to be Ethopian, *Basal* — black rock yielding iron.

Caldera A large basin-shaped volcanic depression many times larger than a vent or crater.

Cave "A naturally occurring cavity... occurring beneath the surface of the Earth or in a cliff or ledge... which is large enough to permit an individual to enter... (U.S. Federal Cave Resources Protection Act of 1988). There is, however, no commonly accepted definition of a cave and a popular definition requires that a cave extends beyond the twilight zone.

Cinder cone Cone-shaped hill built up by ejecta around a volcanic vent.

Crater A bowl-shaped depression, generally in the top of a volcanic cone; diameter seldom more than 300 m.

Ejecta Fragments thrown out by volcanic explosion.

Feldspar A light-coloured mineral composed largely of silicon, oxygen, aluminum and varying proportions of calcium, sodium and potassium.

Fissure eruption Volcanic eruption that occurs along a fissure.

Granite A common, coarsely crystalline igneous rock, consisting chiefly of quartz, feldspar and mica.

Igneous rocks Rocks formed by solidification of magma.

Lava Hot liquid rock at, or close to, the Earth's surface, and its solidified products.

Lava field A more or less well-defined area covered by one or more lava flows.

Lava level lines Horizontal or near-horizontal ledges on cave walls formed by deposition or undercutting when the flow remained at a constant height. At most locations, the width of the ledge is taken to indicate the duration of constant height of the lava river.

Lava tube A conduit formed of hardened lava through which lava flows, or has flowed, to an advancing flow front. In this book the term is restricted to the major conduits to distinguish them from individual segments of tubes usually referred to as **Caves** (see above).

Lava tube cave A segment of a lava tube (see definitions: **"Cave"** and **"Lava tube"** above).

Lava tube system A system of conduits and associated features related to one volcano. The Wall, near Mount Surprise, in the long northwestern flow from Undara Volcano is included in the Undara Lava Tube System because it is considered to reflect the location of the major conduit that allowed this lava flow to become the Earth's longest from a single volcano.

Lavacicle Lava drip formation common in lava tubes, usually triangular in shape; (at Undara, up to 6 cm long).

Magma Molten rock beneath the Earth's surface.

Olivine A mineral, generally green, composed largely of silicon, iron, magnesium and oxygen.

Pahoehoe Lava with smooth or ropy surface. (Pronounced pah ho-eh ho-eh, this Polynesian word, now accepted worldwide, means: "having a satin-like appearance" [Dana, 1891]).

Pele Polynesian goddess of the volcanoes who was thought to cause the volcanoes to erupt when she was angered.

Pele's Hair Volcanic glass spun out into fine brown threads that look like human hair.

Pele's Tears Congealed lava droplets, often found among strands of Pele's Hair.

Phenocryst A crystal that is many times larger than the majority of crystals in an igneous rock.

Phreatic explosion Type of explosion resulting from the interaction of groundwater with hot rocks below the surface of the Earth.

Pyroclastic Made up of fragments ejected from a volcano.

Pyroxene A group of dark-coloured minerals composed mainly of silicon, magnesium, iron, aluminium and oxygen.

Scoria Fragments of porous lava ejected from a volcano.

Spatter cone A cone built by spattery ejecta around a volcanic vent.

Vesicular Having bubble-holes formed by gases.

Viscosity The property of fluids that causes them not to flow easily.

Volcano	A vent through which magma and associated gases and solid material erupt; also applied to a hill or mountain built around such vents. Volcanoes are commonly said to be **active** if they have erupted in historic time, **dormant** if they are quiescent without documented historic activity but are considered likely to erupt in the future, and **extinct** if they have been quiescent for more than about 10,000 years and are considered not likely to erupt again. Many volcanoes, however, are difficult to classify according to these terms: for example, every decade or two, a volcano thought to be extinct returns to activity, sometimes with devastating results.
Note 1:	Worldwide, about 600 volcanoes are classed as active and about two thirds of these lie around the Pacific Ocean.
Note 2:	About 1,300 volcanoes have erupted in the last 10,000 years.
Note 3:	It is only 5,000 years since the last volcanoes erupted in Australia, (Mount Gambier and Mount Shank, South Australia) and some geologists think the future holds potential for renewed activity!

ABBREVIATIONS — USED IN THIS BOOK

A.A.I.P.P.	Associate of the Australian Institute of Professional Photographers.
A.B.C.T.V.	Australian Broadcasting Corporation Television.
A.R.P.S. (Lon.)	Associate of the Royal Photographic Society (London).
A.S.F.	Australian Speleological Federation.
A.U.S.L.I.G.*	Australian Surveying & Land Information Group.*
B.M.R.	Bureau of Mineral Resources. (now Australian Geological Survey Organisation).
C.C.C.	Chillagoe Caving Club.
D.E.H.	Department of Environment and Heritage.
J.C.U.N.Q.	James Cook University of North Queensland.
NASA	National Aeronautical & Space Administration.
Ph.D.	Doctor of Philosophy.
P.U.	Phillips University, Marburg, Germany.
Q.N.P. & W.S.	Queensland National Parks & Wildlife Service.
Q.T.T.C.	Queensland Tourist & Travel Corporation.
R.S.S.	Ravenshoe State School.
T.C. & K.R.G.	Tasmanian Cave and Karst Research Group.
U.C.O.	University of California Observatories.
U.S.	United States.
U.S.A.	United States of America.
U.S.G.S.	United States Geological Survey.
U.W.	University of Wollongong, N.S.W.

* *Aerial photographs supplied by this Group for pages vi and 49 are Crown Copyright and have been reproduced with the permission of the General Manager, Australian Surveying and Land Information Group, Department of Administrative Services Canberra, A.C.T.*

Brief notes to answer some questions often asked about the Earth's interior and volcanic activity

(written with advice from Richard W. Hazlett, Professor of Geology, Pomona College, Claremont, California, U.S.A.).

Even with the deep drilling projects in South Africa and Russia, man has penetrated only to a depth of 3 km, so most of our knowledge of the interior of the Earth comes from the behaviour of earthquake waves and other geophysical and laboratory data.

We know that temperature in mines increases with depth approximately 1°C for every 40 m, the rate of increase becoming smaller with depth. Estimated temperatures at greater depths within the Earth are indicated on the diagram opposite.

...but what causes the rock to melt and rise in a volcano?

Most geologists today think that radioactive decay of elements (in basalt: uranium, thorium and potassium 40) is the primary source of heat that maintains the Earth's internal temperature and that **volcanic activity** acts as a sort of safety valve.

The melting which produces molten rocks occurs in the **upper mantle**, in a zone at the base of the **crust**, the thickness of which varies from 14 km to 60 km in different parts of the Earth.

In the same way that bubbles expand in a soft drink when the lid is lifted, gases dissolved in magma expand as the pressure is reduced. This and the density of the magma being less than that of the surrounding rock, cause the magma to move upward. If it reaches the surface, the gases may cause the lava to fountain.

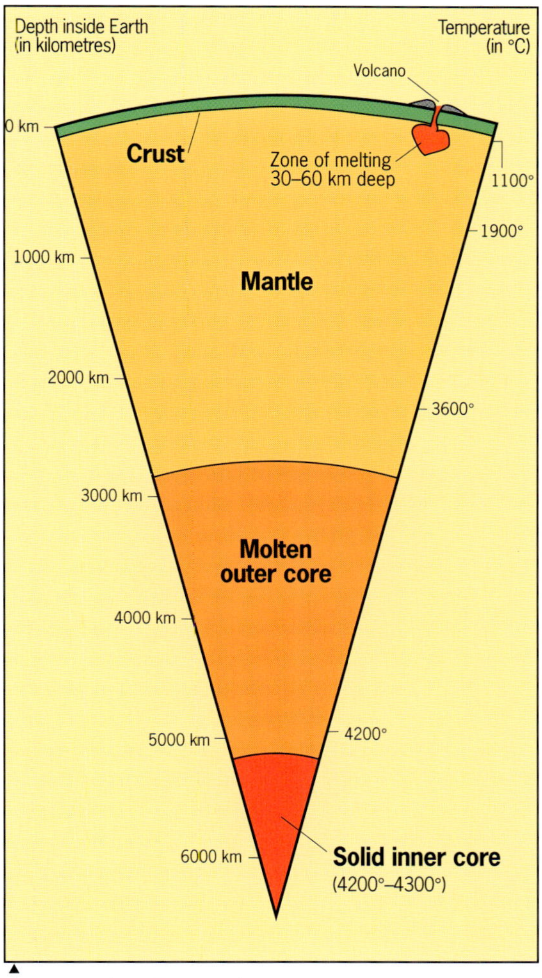

Segment of the Earth to show the area where basaltic lavas are believed to originate and the estimated temperatures at various depths within the Earth.

Figures in the above diagram are from R.W. Hazlett in Decker, Robert and Barbara, 1991.

Chemical composition and its effect on type of eruption

From Macdonald and Hubbard, 1989

Most of the world's volcanoes erupt more violently than those in Hawaii. The reason is that Hawaiian lavas are less viscous (more fluid) and have a lower gas content.

The three main factors that determine the viscosity of a lava are:

- temperature
- chemical composition
- gas content

When the percentage of silica in a lava is high, it is more viscous and it is difficult for the gases in it to escape gently. The pressure of gas therefore builds up until a violent explosive eruption occurs. Below are tabulated analyses of average Hawaiian lava and that of the lava from the explosive eruption of Mt. St. Helens in 1980. They provide an interesting example of what the difference in SiO_2 percentage can mean.

One analysis from the long northwestern flow [see map, page (1)] from **Undara Volcano** is also tabulated. The similarity between the SiO_2 percentage in the Undara analysis and that in the average Hawaiian lava, coupled with other data, allows the conclusion that eruptions from the Undara Volcano were non-explosive, similar to many Hawaiian eruptions in historic time.

	Hawaiian Islands #	Mt. St. Helens (1980)	Undara *
Silicon .... SiO_2	48.4	63.5	48.9
Titanium .... TiO_2	2.8	0.6	1.8
Aluminium .... Al_2O_3	13.2	17.6	15.2
Iron .... FeO	11.2	4.2	10.5
Magnesium...................... MgO	9.7	2.0	8.6
Calcium .... CaO	10.3	5.2	9.7
Sodium .... Na_2O	2.4	4.6	3.9
Potassium .... K_2O	0.6	1.3	1.8
Other Oxides	1.4	1.0	—

Table of average chemical composition *of lavas expressed as weight percent oxides from the Hawaiian Islands, Mt. St. Helens, U.S.A., explosive eruption (1980) and one analysis from an Undara flow.*

Average of 53 analyses of olivine basalts.
* Host rock, Barkers Cave entrance; analysis: T.J. Griffin.

Atkinsons' first map and sections, Undara Lava Tube System.

◂ *Tom Atkinson, 1950–1988, — A talented pilot and an observant bushman with an inquiring mind. We remember with affection and gratitude his skill in the air and on the ground, his interest and support in our early investigations.*

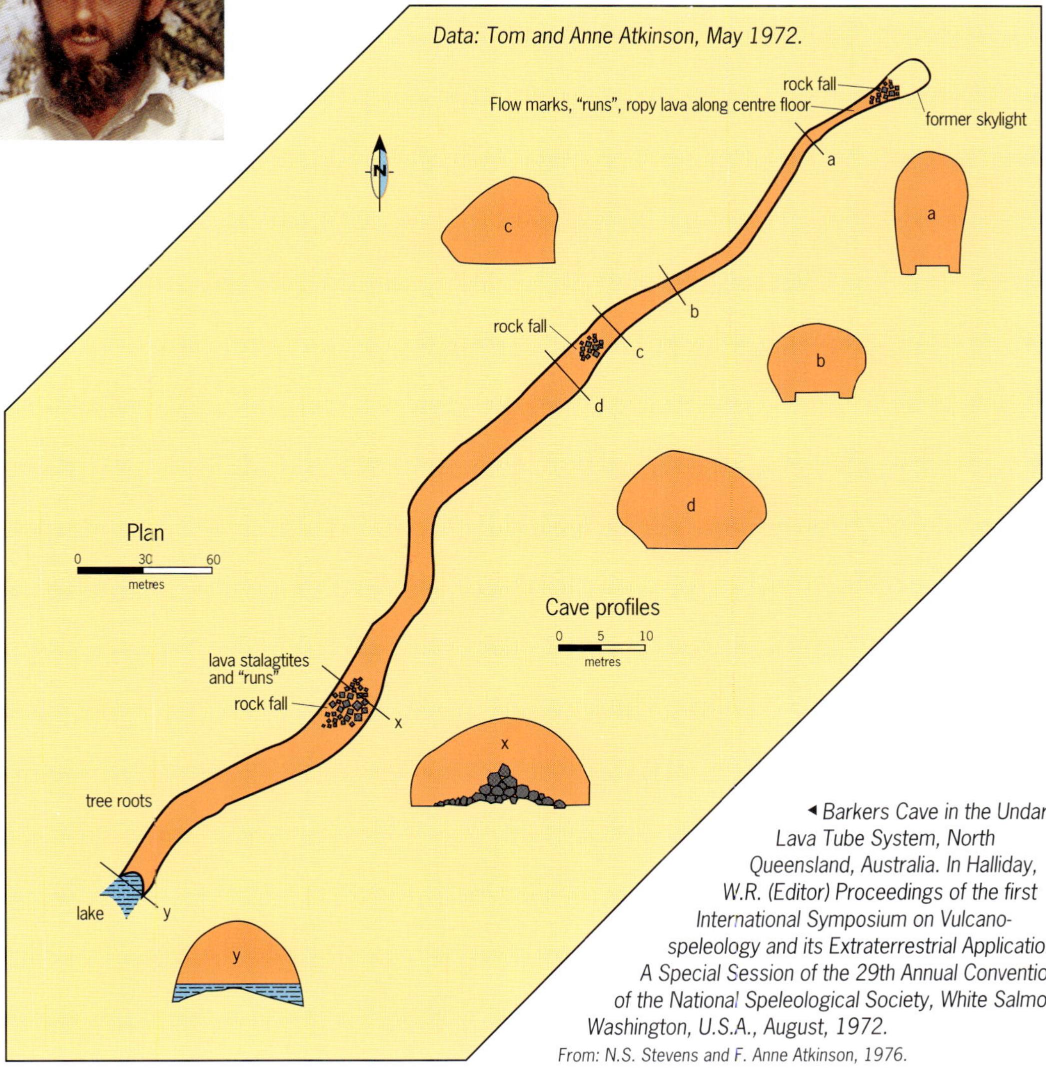

◂ *Barkers Cave in the Undara Lava Tube System, North Queensland, Australia. In Halliday, W.R. (Editor) Proceedings of the first International Symposium on Vulcanospeleology and its Extraterrestrial Applications. A Special Session of the 29th Annual Convention of the National Speleological Society, White Salmon, Washington, U.S.A., August, 1972.*

From: N.S. Stevens and F. Anne Atkinson, 1976.

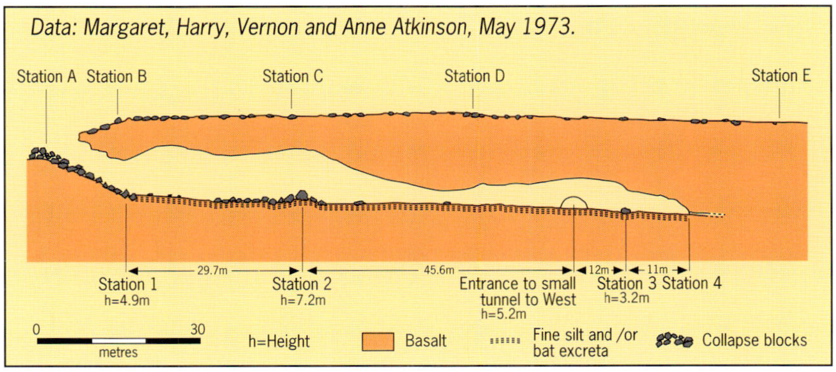

▴
First Logitudinal Profile, Undara Lava Tube System: Atkinson Cave.

Photos from our album — family, friends and others involved

▲ *1968, Brisbane members of National Parks Association of Queensland in Barkers Cave — their first visit to Undara.*
Photo: N. Traves.

▲ *Norm Traves and family in Collins Road cave.*
Photo: M. Spaulding.

▲ *Russell Dunmall from R.S.S., a valued student assistant.*

▲ *1969 Traves family introduced Atkinson family to Undara caves; photographed in Barkers Cave.*
Photo: N. Traves

◂ *1973. Homeward bound! Chief field assistant, Darcy Day and family on truck ready for the trip home. In an area with no surface water and no roads, the length of our field trips was governed by our management of the 200 litre drum of water we carried.*

▲
Dr. Ivan Lester, F.R.C.S., F.R.A.C.S., the person whose advice initially led us to accept the challenge of research at Undara. In 1972, after he carried out successful cancer surgery on Anne, he insisted that she find an absorbing interest far removed from routine country life. Quite by chance she heard, soon after, that fieldwork was required urgently at nearby Undara for a contribution to the first International Symposium on Vulcanospeleology, being convened at the request of NASA, to collect data of the world's lava tubes.

▲
1972, Vernon and Anne as they set out on a long, long journey.
Photos: P.J. Stephenson, J.C.U.N.Q.

▲
Lessons finished — we all go underground.

▲
For the youngest field assistant, Matthew, correspondence lessons were continued… with the reward (?) — a trip underground.

▲
Though it gave him no ambition to become a geologist, Matthew seemed not to resent the fatigue from the long hours underground.

▲
Finally back to camp.

▲
Above right and left: July 1973: Trig station, Mauna Loa, 4168m. Right: view from the rim, across the caldera — not everyone's choice of a picnic venue, but for Anne — a school days' ambition realised! Photos: Vernon Atkinson.

Operation Raleigh at Undara 1988

▲
Building their Undara camp.
Photo: Brian Furniss, O.R.

▲
Every settlement needs a name…
Photo: Brian Furniss, O.R.

▲
Getting oriented from Barkers Knob.
Photo: Brian Furniss, O.R.

▲
Leader Brian Furniss abseiling into cave.
Photo: Richard Murray, The Australian Newspaper.

◂ *Members surveying in a cave that they had just discovered.*
Photo: Brian Furniss, O.R.

▲
A fairytale bride came through Lacy Scrub.

The first Undara Wedding

3rd September,1993 Bramwell John Collins married Sarah Joanna Fulford-Kirby

"Through dim lit aisles and halls of green..."*Strauss.*

...to the haunting sound of the didgeridoo. ▸

To meet her bridegroom in "The Archway" — a 190,000 year old cathedral.
Photos: Mark Johnson, by courtesy Bram and Sarah Collins.
▼

▸APPENDIX V

Tabulation of some of the world's long lava caves. All are in lava flows that are younger, or much younger, than Undara's flows. Lengths are from current records of the National Speleological Society of U.S.A., except where indicated.

As far as can be ascertained the following list gives the longest segment in the system. However, it should be remembered that the definition of a segment is somewhat arbitary.

Country	Length metres	Name
* U.S.A. — Hawaii	>25,000	Kazamura
South Korea	13,399	Mangung-gul
Canary Islands	12,499	Cueva del Viento
South Korea	11,748	Bilremos-gul
Kenya	11,152	Leviathan
U.S.A. — Hawaii	7,100	Ainahou Ranch
U.S.A. — California	6,660	Labyrinth
# Canary Islands	6,315	Cueva de Don Justo
U.S.A. — Hawaii	6,263	John Martin
U.S.A. — Hawaii	6,067	Olaa
U.S.A. — Washington	3,904	Ape
U.S.A. — Utah	3,674	Duck Creek
U.S.A. — Hawaii	3,400	Offal
• China	2,000	Laoheishan
◆ Iceland	1,970	Surtshellir
	1,585	Viogelmir
	1,520	Stephenshellir

* Personal communication, 1995, U.S.G.S. Hawaii Volcano Observatory, Hawaii.

• Personal communication, Dr. Lu Longwen, Director, Wudalianchi Earthquake –Volcano Observatory, China.

◆ Icelandic Caving Journal, 1990.

J.J. Hernández, Medina, A.L. & Izquierdo, I., 1991: in Rea, G. Thomas, (Editor) Proceedings of the 6th International Symposium on Vulcanospeleology, Hawaii.

◂ *Blind centipede from Bayliss Cave, Undara Lava Tube System, a member of the family scutigeridae and one of the world's largest terrestrial, cave-adapted invertebrates. Note the very long antennae. Extra development of other senses compensates for its lack of sight. Body length of this species is up to 8cm.* Photo: F. Howarth, Bishop Museum, Honolulu.

▸APPENDIX VI

MAMMALS recorded within and near the Undara Volcanic National Park

We are not attempting to provide comprehensive lists of the interesting fauna and flora recorded in the **Undara** area. We hope that, at some not too distant date experts in these fields will present these in an illustrated companion volume. In the meantime current lists can be obtained from the office, Undara Lava Lodge or by writing to:

Department of Environment and Heritage,
P.O. Box 2066, CAIRNS, Q 4870, AUSTRALIA.

For the interest of our readers, with advice from L. Little and M. Godwin, Department of Environment and Heritage, Cairns, and S. Van Dyck, Curator, (Vertebrates), Queensland Museum, we include here a list of mammals identified to date.

Common Name	**Scientific Name**
Agile wallaby*	*Macropus agilis*
Antilopine wallaroo	*Macropus antilopinus*
Black-striped wallaby	*Macropus dorsalis*
Common wallaroo	*Macropus robustus*
Eastern grey kangaroo	*Macropus giganteus*
Mareeba rock wallaby	*Petrogale mareeba*
Red-legged pademelon	*Thylogale stigmatica*
Rufous bettong	*Aepyprymnus rufescens*
Spectacled hare wallaby	*Lagorchestes conspicillatus*
Swamp wallaby	*Wallabia bicolor*
Whiptail wallaby	*Macropus parryi*
Greater glider	*Petauroides volans*
Squirrel glider	*Petaurus norfolcensis*
Common brushtail	*Trichosurus vulpecula*
Koala*	*Phascolarctos cinereus*
Common bentwing bat	*Miniopterus schreibersii*
Common sheathtail bat	*Taphozous georgianus*
Diadem bat	*Hipposideros diadema*
Dusky horseshoe bat	*Hipposideros ater*
Eastern horseshoe bat	*Rhinolophus megaphyllus*
Ghost bat*	*Macroderma gigas*
Gould's wattled bat	*Chalinolobus gouldii*
Little bentwing bat	*Miniopterus australis*
Little cave bat	*Vespadelus troughtoni* (formerly identified as *Eptesicus pumulis*)
Little red flying fox	*Pteropus scapulatus*
Echidna	*Tachyglossus aculeatus*
Long-tailed planigale	*Planigale ingrami*
Northern quoll	*Dasyurus hallucatus*
Short nosed bandicoot	*Isoodon macrourus*
Dingo	*Canis familiaris*

* *very rare.*

Photo: O. Frank

"The world will never starve for want of wonders,
but only for the want of wonder". *G.K. Chesterton.*